A Natural History of Natural Theology

A Natural History of Natural Theology

The Cognitive Science of Theology and Philosophy of Religion

Helen De Cruz and Johan De Smedt

The MIT Press
Cambridge, Massachusetts
London, England

This book was set in Stone Sans Std and Stone Serif Std by Toppan Best-set Premedia Limited, Hong Kong.

Library of Congress Cataloging-in-Publication Data

De Cruz, Helen, 1978–.
A natural history of natural theology : the cognitive science of theology and philosophy of religion / Helen De Cruz and Johan De Smedt.
  pages   cm
Includes bibliographical references and index.
ISBN 978-0-262-02854-7 (hardcover : alk. paper), 978-0-262-55245-5 (paperback)
1. Natural theology. 2. God—Proof. 3. Cognition. 4. Religion—Philosophy.
I. Title.
BL183.D4   2014
210—dc23
2014017234

For Aliénor and Gabriel

Contents

Acknowledgments

The writing of this book has been financially supported by a fellowship from the University of Oklahoma's Oxford Templeton Fellows Program, the Research Foundation Flanders, the British Academy, and Ghent University. We particularly want to thank Linda Zagzebski, director of the Oklahoma program, for providing the opportunity to conduct research and study at Oxford and to pursue this project, and Tim Mawson, Dave Leal, Steve Clarke, and Rob Byer for help with the practical aspects of our stay.

Chapters 4 and 5 incorporate material (significantly revised and expanded) from two journal articles, respectively Helen De Cruz and Johan De Smedt, "Paley's iPod: The Cognitive Basis of the Design Argument within Natural Theology," *Zygon: Journal of Religion and Science* 45 (2010a): 665–684, and Johan De Smedt and Helen De Cruz, "The Cognitive Appeal of the Cosmological Argument," *Method and Theory in the Study of Religion* 23 (2011): 103–122. We thank Wiley-Blackwell and Brill for granting permission to reuse this material.

Some of the chapters in this book were presented and discussed at work-in-progress seminars at Oxford's philosophy faculty, the Ian Ramsey Centre for Science and Religion, and the Joseph Butler Society. We are indebted to faculty members, visiting scholars, and graduate students at Oxford who have discussed our papers, and who have read and commented upon individual chapters: John Cottingham, Joseph Diekemper, Klaas Kraay, Brian Leftow, Jennifer Nagel, Andrew Pinsent, Emily Reed Burdett, Wes Skolits, Richard Swinburne, Tobias Tan, Olli-Pekka Vainio, Aku Visala, and William Wood.

We are also grateful to Mitch Hodge and Ryan Nichols, who have read drafts of some chapters of this book. We want to thank the three reviewers of the MIT Press who provided very helpful comments, namely Steve Horst and Kelly James Clark (who revealed their identity to us), and an additional anonymous referee. We also thank Philip Laughlin, Judith Feldmann, and Christopher Eyer at the MIT Press for their invaluable help in the production process of this book.

Finally, we would like to thank our children, Aliénor, who witnessed the book taking shape in the course of three years at Oxford, and Gabriel, who was born when the writing was in progress. Many thanks for their patience with their scholarly parents during the writing process.

Introduction

Theists in the Abrahamic monotheistic traditions believe in the existence of an omniscient, omnipresent, eternal, morally perfect being who created the world and sustains its existence continuously. How can we know whether such a being really exists, and if he exists, can we find out what he is like? These two questions, the existence and the attributes of God, form the subject matter of natural theology, the discipline that seeks to obtain knowledge of God by relying on reason and experience of the world. The term *natural theology* covers several fields of inquiry, including the formulation of natural theological arguments, the search for the natural knowledge of God that humans possess, and the conceptualization of nature, including an interpretation of the natural sciences, within a theistic framework.

In its most-often used sense, natural theology is understood as the formulation of arguments for or against the existence of God or some of his attributes. This meaning is so common that the term is often used synonymously for such arguments. Examples include the cosmological argument, which starts from the observation that the world exists and infers from this a necessary being on which the world is causally dependent; the moral argument, which takes the objectivity of moral norms as evidence for the existence of a morally perfect God who instilled these norms; and the argument from design, which takes the complexity and fine-tuning of the cosmos or some of its features as evidence for a divine designer. There are also natural theological arguments against the existence of God, such as the argument from evil, which takes the existence of evil as evidence against an omnipotent and morally perfect God.

Arguments in natural theology rely to an important extent on intuitions and inferences that seem natural to us. Such intuitions occur spontaneously even to nonphilosophers in their reflective moments. It does not require years of philosophical or theological training to infer a designer from the appearance of design and purposiveness in the natural world. Likewise, it seems natural to ask why there is something rather than nothing: how do we account for the existence of the world, which seems so contingent and almost accidental to us? One can be struck by the objective force that moral norms seem to have, and one can infer that there are indeed absolute, culture-independent moral truths. From this, it seems plausible and natural to assume the existence of someone who has inscribed those moral intuitions in our minds. Watching a dramatic formation of clouds on an evening walk in the meadows can elicit an argument from beauty: aesthetic experiences like these may compellingly point to the existence of God. These intuitions about design, existence, agency, morality, and beauty have been formulated over and over in history and in disparate religious traditions, and they continue to be relevant today.

Yet, in spite of their intuitive appeal, the premises that underlie natural theological arguments have often been criticized. More than a century before Darwin (1859) and Wallace (1858) developed the theory of natural selection as a naturalistic mechanism to explain design in nature, authors like David Hume criticized the argument from design. Given our current detailed understanding of how natural selection operates, the argument from design has lost a lot of its initial cogency. In response to various versions of the cosmological argument, critics have pointed out that it may well be that the existence of the universe is a brute fact that simply does not require an explanation (Grünbaum 2000), or even more mysteriously, that the universe caused itself to exist (Smith 1999).

Intuitions are peculiar things: for those who hold them, they are obvious, sometimes self-evident; others will find those same intuitions not obvious at all, far-fetched, or easy to undermine. This is also the case for the intuitions that underlie natural theological arguments. A possible source for disagreement may be that the plausibility of these intuitions is mediated by prior beliefs and expectations. For instance, whether one believes that all things need a cause for their existence may be mediated by one's prior beliefs about the existence of God.

Thus far, we have identified two puzzling features of natural theological arguments: they rest on intuitions that are untutored and, to some, appear obvious and self-evident. At the same time, there has been and continues to be disagreement about the validity of these intuitions. Some philosophers (Alston 1991; Nagel 2012) have argued that we should treat intuitions that underlie philosophy—and by extension natural theology—as instances of ordinary reasoning. They argue that there are important continuities between intuitions and cognitive faculties such as perception. Thomas Reid (1764) proposed that some intuitions, which he termed commonsense beliefs, are naturally produced by our cognitive faculties. These include the belief that perception is reliable, that the world really exists, that other people have minds, and that induction is generally trustworthy. Rejecting such beliefs would seriously compromise our ability to interact with the world: we would be utterly incapable of relating to other people, unable to make generalizations, and constantly have to distrust our senses. Therefore, Reid concluded, it is unreasonable to doubt commonsense beliefs.

Not all intuitions belong to the realm of common sense but can be quite remote from our everyday experience. A physicist may have intuitions about the plausibility of string theory; a philosopher may have intuitions about whether or not human free will is compatible with determinism. Such intuitions seem to be compelling to those who hold them. As Peter van Inwagen (1999, 274) puts it, "I believe that I am justified in believing [these intuitions]. And I am confident that I am right. But how can I take these positions? I don't know. That is itself a philosophical question, and I have no firm opinion about its correct answer. I suppose my best guess is that I enjoy some sort of philosophical insight."

Little work has been done on the psychological origins of our intuitions. There is a growing body of experimental philosophical research on the content of intuitions (see Alexander 2012 for a review) and the extent to which they vary between different groups, such as between cultures or genders. However, these authors seem to be mainly interested in pinpointing what intuitions ordinary people have, for instance, whether laypeople hold that knowledge is justified true belief. They have not yet explored where such intuitions originate. Why do we spontaneously intuit that order and design require a designer? What is peculiar about aesthetic experience that it leads some people to infer the existence of God? The main aim of this book is to examine the cognitive origins of these and other natural theological intuitions. We will see that many seemingly arcane natural theological intuitions are psychologically akin to more universally held, early-developed, commonsense intuitions.

In recent years, cognitive scientists have begun to investigate the psychological underpinnings of religious beliefs. Using evidence from fields like anthropology, developmental psychology, and neuroscience, they have convincingly argued that religion relies on normal human cognitive faculties. Religious beliefs arise early and spontaneously in development, without explicit instruction. They exhibit cross-cultural similarities such as afterlife beliefs and a belief in supernatural agents. Such beliefs are examined in terms of ordinary cognitive biases and constraints. Philosophers are exploring the implications of this research for the justification of religious beliefs. Can the rationality of religious beliefs be maintained in light of the fact that they are the result of these cognitive processes? In these discussions, cognitive scientists of religion have focused on what may be termed "folk" religious beliefs and practices rather than on theological reasoning. In a way, this makes sense because theological reflection constitutes only a tiny part of religious experience since it is mostly performed by specialists in highly institutionalized contexts, such as universities and monasteries. Through this neglect of theology, cognitive scientists of religion have not yet addressed the question of whether theological reflection is continuous with everyday cognitive processes—they have naturalized religion, but they have yet to naturalize theology.

Traditionally, cognitive scientists have argued for a large cognitive divide between folk religion and theology. They view folk religious beliefs as cognitively natural, but theology as unnatural (McCauley 2011). For example, Justin Barrett (2004) asserts that people explicitly claim to accept official theology (theological correctness) but that they implicitly hold theologically incorrect beliefs—they distort their beliefs about God in such a way that these match their intuitive expectations about ordinary human beings. This received opinion on the unnaturalness of theology does not sit well with the observation that intuitions that underlie natural theological arguments are obvious, self-evident, and compelling. This observation provides a starting point for an empirically informed study of the cognitive basis of natural theology. Using evidence and theories from the cognitive science of religion and cognate disciplines—such as evolutionary ethics, evolutionary aesthetics, and the cognitive science of testimony—we aim to show that natural theological arguments and inferences rely to an important extent on intuitions that arise spontaneously and early in development and that are a stable part of human cognition.

If natural theology rests on cognitively natural foundations, this has implications for the cogency and viability of natural theological arguments.

There is an established tradition in theology with Aquinas and Calvin, harkening back to Paul's letter to the Romans, that argues that humans possess a natural knowledge of God. From this perspective, the intuitions that underlie natural theological arguments may have theological significance, as they could be an integral part of this natural knowledge. However, natural theological arguments do not provide knowledge if the intuitions that lead philosophers and theologians to formulate natural theological arguments are spurious or if they are unjustified in applying them to metaphysical questions. Throughout this book and in chapter 9 in particular, we will investigate to what extent the cognitive origins of natural theological intuitions undermine or strengthen these arguments.

This book is structured as follows: chapter 1 introduces the main themes through David Hume's two questions on religion: its foundation in reason (natural theology) and its origin in human nature (what was then termed "natural history"). We look at natural theology and cognitive science of religion as two disciplines that attempt to answer these questions. Chapter 2 provides theoretical background on the cognitive science of religion and its assumptions. This chapter will also clarify the use of the terms *natural* and *intuitive* in cognitive science and philosophy of religion. Chapter 3 looks at intuitions that underlie natural theological reflections on divine attributes, in particular omniscience. We consider experimental results on how children and adults think about the knowledge of others, and we relate these to discussions in natural theology. Chapters 4 through 8 present five well-known natural theological arguments for theism: the argument from design, the cosmological argument, the moral argument, the argument from beauty, and the argument from miracles. Each of these chapters will investigate the cognitive underpinnings of the intuitions that underlie these arguments. We will consider why there is disagreement about them in spite of their intuitive appeal. We will also explore whether theists and nontheists are reasonable in holding the beliefs they do in light of the results of cognitive science. Chapter 9 concludes by exploring evolutionary debunking arguments against religious belief and, more specifically, against natural theology. We argue that intuitions that fuel natural theological argumentation are not immune to causal, naturalistic accounts, as these intuitions result from the workings of normal human cognitive processes. As a result, a neat separation between the psychological origins of religion and its foundation in reason cannot be maintained.

1   Natural Theology and Natural History

Natural theology and the cognitive science of religion are two fields that seem to have little in common. Natural theology is concerned with obtaining knowledge of God through experience and reasoning, whereas the cognitive science of religion studies religious beliefs and practices through naturalistic, scientific methods. This chapter shows how key questions in both disciplines are nevertheless interrelated and how results from the one can have implications for the other. Throughout its history, natural theology has always maintained a dialogue with the sciences, which makes cognitive science of religion potentially interesting to natural theologians.

Two Questions about Religion

As every enquiry, which regards religion, is of the utmost importance, there are two questions in particular, which challenge our principal attention, to wit, that concerning its foundation in reason, and that concerning its origin in human nature.
—David Hume (1757, 1)

In *The Natural History of Religion*, David Hume (1757) posed two central questions about religion: is religious belief reasonable, and what is it about human nature that makes people hold such beliefs? The first question is the subject matter of natural theology, as it examines on what grounds one can rationally hold particular religious beliefs. The second question concerns the causal origin of religious beliefs. In the eighteenth century, natural history was the intellectual endeavor of uniting empirical facts through theoretical, explanatory frameworks. Traditionally, the natural history of religion and natural theology have been treated as separate questions. Hume's *Natural History of Religion* traces the historical development of religion and its roots in human psychology (the second question), whereas his posthumously published *Dialogues Concerning Natural Religion* (1779) considers the cogency of natural theological arguments (the first question).

As Hume already adumbrated, and as we will argue in this book, the two questions on religion are intimately connected. One of our main claims is that an understanding of the foundation of religion in reason requires a better insight into how reason and causal origin relate, in particular where intuitions in natural theology originate. Hume proposed that such intuitions come to mind simply because they are plain to us: "The whole frame of nature bespeaks an intelligent author; and no rational enquirer can, after serious reflexion, suspend his belief a moment with regard to the primary principles of genuine Theism and Religion" (Hume 1757, 1). Nevertheless, he expressed skepticism about the design argument in *Dialogues*, suggesting that intuitions that underlie natural theology are not self-evident after all. If these intuitions are not self-evident, we can ask whether the inference to theism through reason and observation is warranted.

The natural history of religion and natural theology also intersect in the extent to which causal explanations of religion, offered by the sciences, have implications for the reasonableness of religious beliefs. Simply put, does a naturalistic (evolutionary, cognitive) explanation for religious beliefs debunk these beliefs? The term "debunking" is commonly used in a strong sense, namely "to refute," to prove that something is false. However, philosophers usually employ the term in a more restricted way, in the sense of "to explain away." Debunking arguments examine the causal history of a particular belief in a way that undermines that belief. For instance, suppose one can explain religious belief as an adaptation that allows us to increase cooperative behavior: through religious beliefs and the accompanying costly religious commitments such as dress codes and food restrictions, humans can reliably signal that they are part of the in-group, which, as a result, facilitates human cooperation (Sosis and Alcorta 2003). Nothing in this explanation makes any causal reference to God. So, while the existence of God is not conclusively disproved, a debunker could argue that this explanation still undermines the rationality of theism. We will look more closely at such arguments in chapter 9.

In recent years, a growing number of philosophers of religion have focused on the question of whether or not evolutionary and cognitive explanations of religious belief can support debunking arguments against God's existence (e.g., the essays collected in Schloss and Murray 2009). Perhaps unsurprisingly, theistic authors have argued that such explanations do not undermine theism (e.g., Visala 2011), or even that they provide support for theism (e.g., Barrett 2011), whereas nontheists suspect that these approaches debunk or are at least problematic for theism (e.g., Bering 2011). In this book, we will concentrate mostly on the first type of intersection between natural history and natural theology, namely what the sciences of the mind can tell us about intuitions that underlie natural theology. However, as we will see, understanding better how natural theology relies on stable human cognitive capacities will have implications for the rationality of natural theological arguments as well.

There is also a broader metaphysical consideration of why we can expect natural theology and natural history of religion to converge. Those who assume that metaphysical naturalism is the correct worldview, that is, that there is no God, would still need to explain why it is that intuitions about design, causality, agency, beauty, morality, and testimony persistently lead people to infer the existence of God. Why are design arguments compelling if there is no cosmic designer? Why do humans draw a connection between human moral intuitions and divine moral commands if there is no God who instilled these moral intuitions in us? A plausible explanation is that such inferences draw upon deep-seated human psychological dispositions. Hume recognized this in his *Dialogues*, where Cleanthes (the proponent of natural theological arguments for theism) queries Philo's unflinching use of reason to attack natural theological arguments. Philo responds as follows:

I must confess, replied Philo, that I am less cautious on the subject of natural religion than on any other; both because I know that I can never, on that head, corrupt the principles of any man of common sense; and because no one, I am confident, in whose eyes I appear a man of common sense, will ever mistake my intentions. ... A purpose, an intention, a design, strikes every where the most careless, the most stupid thinker; and no man can be so hardened in absurd systems, as at all times to reject it. (Hume 1779, 228)

This passage, commonly referred to as Philo's confession, has received much attention in recent Hume scholarship (e.g., Goodnick 2012; Hardy 2012). One way to interpret it is that Hume thought that humans spontaneously see design, which leads them to posit the existence of a designer. As we shall see in chapter 4, metaphysical naturalists today can explain this propensity as the result of natural selection. If metaphysical naturalism is correct, inferring God from design, causality, moral judgments, beauty, and testimony are mistakes, misapplications of intuitions that have evolved to function in different domains.

By contrast, under a theistic worldview, natural theological intuitions can be interpreted quite differently, for under such a view (recently defended by Evans 2010), it seems plausible that God made the universe and designed the human mind in such a way that traces of his workmanship can be clearly discerned. Theists often invoke scriptural support for the view that God made his existence knowable not only to trained scientists and theologians, but to everyone who observes the world.

The heavens are telling the glory of God; and the firmament
    proclaims his handiwork.
Day to day pours forth speech, and night to night declares
    knowledge.
There is no speech, nor are there words; their voice is not heard;
    yet their voice goes out through all the earth, and their words to
    the end of the world. (Psalm 19:1–4)[1]

For what can be known about God is plain to them, because God has shown it to them. Ever since the creation of the world his eternal power and divine nature, invisible though they are, have been understood and seen through the things he has made. So they are without excuse; for though they knew God, they did not honor him as God or give thanks to him, but they became futile in their thinking, and their senseless minds were darkened. (Romans 1:19–21)

Appeals to scripture fall outside of the scope of natural theology, but they are often used to justify its practice (e.g., Sudduth 1995). Theologians like John Calvin ([1559] 1960, book 1) have interpreted these scriptural passages as indicating that our ability to recognize God's handiwork (i.e., to engage in natural theology) is reliably connected to God's existence. Interestingly, they treat natural theology not as an arcane endeavor that is remote from everyday reasoning, but propose that discerning God's handiwork is something that we do spontaneously and that is not dependent on a prior assumption of theism. After all, if Paul believed that this discernment was dependent on prior theistic belief, he would not have thought unbelievers were without excuse.

The metaphysical worldview one adopts has significant ramifications for one's conceptualization of the intersection between cognitive science and natural theology. If metaphysical naturalism is right, the intuitions that underlie natural theology are incorrect; if theism is right, they are correct. Nevertheless, investigating the cognitive basis of natural theological arguments can be pursued largely independently from either of these worldviews. Even if one adopts a theistic worldview, according to which God designed the world and human cognition in such a way as to make himself knowable, it is still worthwhile to have a detailed causal account of this. Likewise, within a nontheistic framework, it is interesting to explain the enduring intuitive appeal of natural theological arguments by looking at stable features of human cognition.

Throughout this book, we will take moderate naturalism as a methodological assumption. Thus, when investigating the cognitive basis of intuitions in natural theology, we will not adopt metaphysical naturalism, which holds that there are no supernatural entities, nor will we assume a metaphysical theism, which takes the existence of God as a given. Moderate naturalism is neutral with respect to metaphysical assumptions. Alvin Goldman defines it as follows:

Moderate naturalism

(A) All epistemic warrant or justification is a function of the psychological (perhaps computational) processes that produce or preserve belief.[2]
(B) The epistemological enterprise needs appropriate help from science, especially the science of the mind. (Goldman 1999, 3)

Our chief help in our investigation of the cognitive basis of natural theology will be the cognitive science of religion, which investigates the cognitive processes that underlie religious beliefs. Because moderate naturalism is not committed to either theism or metaphysical naturalism, the reasonableness of intuitions that underlie natural theology cannot be assessed only by considering their psychological origins. For this, we will require additional assumptions about the relationship between the psychological origin of a belief and its justification. Throughout this book we will consider these assumptions as well.

What Is Natural Theology?

Natural theology is a branch of theology that examines the existence and attributes of God (or the gods, in polytheistic traditions) without reliance on special revelation. As a starting point, it takes the world of our ordinary experience (i.e., general revelation), and considers this in a systematic, reasoned way to link it to another, asserted reality. It is often contrasted with revealed theology, which depends on revelation through scripture and religious experience. Natural theologians set out to learn something about God, for example, whether or not there is a God, and what properties he has (the divine attributes), using ordinary experience and reasoning. In this broad sense, natural theology is also known as philosophical theology.

Within natural theology, several approaches can be outlined. The best-known and best-developed strand of natural theology is concerned with formulating arguments for or against the existence of God, such as the cosmological and moral arguments. A second strand focuses less on the world and more on the nature and structure of human cognition, in particular on the nature of human desires and beliefs. It regards the human mind as oriented toward God. The common consent argument (see chapter 9) and C. S. Lewis's argument from desire (see chapter 7) are examples of this. A third project is concerned with outlining a theology of nature, which does not set out to infer the existence of God but, rather, aims to understand and interpret the natural world under the assumption of his existence. For instance, the molecular biologist Kenneth Miller (2007), the geneticist Francis Collins (2007), and the theologian John Haught (2000) paint theistic pictures that take into account evolutionary biology.

Natural theology is not restricted to particular religious traditions. Although many of its practitioners were and are monotheists, some polytheistic religions, such as those from India and ancient Greece, also had flourishing natural theologies. Since the seventeenth century, there has been an established tradition of atheist and agnostic natural theology as well (e.g., Hume, Voltaire) that continues to this day, especially with work on the problem of evil and on divine hiddenness (e.g., Trakakis, Schellenberg).

What constitutes the enduring appeal of natural theology across times and cultures? How can it be linked to religious beliefs and practices? Some authors (e.g., Philipse 2012) have argued that one should engage in natural theology before accepting any religious belief because one must first establish through argument and conceptual analysis whether particular religious beliefs are reasonable before one can even begin to adopt or reject them. Whether individual believers are under some form of epistemic obligation to ground their beliefs in natural theology is a matter of enduring philosophical debate. Reformed epistemologists argue that one is entitled to hold religious beliefs in the absence of any positive arguments in their favor (e.g., Plantinga 1993), whereas other authors propose that faith is an attitude that is cognitively and epistemically distinct from belief (e.g., Audi 2011).

Regardless of that, few people are religious believers or atheists because they are swayed by natural theological arguments. Most religious practice does not engage with and is independent from natural theological considerations. Yet natural theology has, as Allister McGrath (2011, 18) puts it, "a persistent habit of returning, even when its death notice has been extensively and repeatedly published." Today, a large part of philosophy of religion (especially in the analytic tradition) focuses on natural theological themes such as arguments for the existence of God, whether God's existence is compatible with moral and natural evil, and the relationship between human free will and divine omniscience. The idea that nature can point us toward some transcendent reality remains a fascinating one. Public debates and lectures on natural theology gather widespread attention, for example, between the former Anglican archbishop Rowan Williams and the evolutionary biologist Richard Dawkins.[3] Natural theological arguments are discussed on numerous atheist and theist blogs and in other popular media. This public attention is remarkable, especially if we consider that natural theological arguments do not play a major role in the motivation to adopt or maintain religious belief.

The history of natural theology in the West can be traced to ancient Greek philosophy, in particular to the argument from design. In spite of the widespread appeal to teleology (purposiveness) to infer the existence of the gods, formal arguments from design remained relatively rare throughout ancient philosophy. The earliest Western design argument is attributed to Socrates, recorded in Xenophon's ([4th c. BCE] 1997) *Memorabilia* (55–63; 299–309). Socrates argued that there are two possible explanations for how living things are brought into being: design and chance. It is clear that living beings exhibit teleology. Products of chance typically do not exhibit teleology, unlike products of design, which often do. Therefore, that which produced living beings did so by design. Later on, Roman Stoic philosophers such as Cicero argued that the orderliness of the world provided strong evidence for the existence of the gods (see chapter 4). In his review of creationism in antiquity, David Sedley (2007) attributes the emergence of natural theological arguments in the ancient world to the rise of atomism, a competing, nontheistic worldview. According to atomism, as espoused by Democritus and Lucretius, the apparent design features came about by chance: long ago, chance produced a variety of life forms, most of them unviable but some of them viable. The living forms today are descendants of these fortuitous results of a blind trial-and-error process. Atomism provided a plausible alternative to creationism. As a result, theism was no longer the default position, and theists such as Socrates and Cicero felt obliged to provide explicit arguments that outlined a positive case for it.

This dialectical function of natural theology can also be discerned in non-Western traditions (see De Cruz 2014 for a detailed treatment). For instance, Hindu natural theologians formulated design and cosmological arguments as a reaction to nontheistic philosophical schools that included materialism, atomism, some varieties of Buddhism, and Sāmkhya, an atheist form of evolutionism. These views threatened a developing creationist form of Hinduism, according to which Brahman is the personal creator of the world. Although theistic authors like Śaṅkara and Udayana thought that scripture was authoritative and superior to reason, they still felt compelled to go against these schools by using natural theological arguments because proponents of these movements did not accept their interpretation of scripture or even rejected scripture altogether (Brown 2012, chapters 2–4).

Western natural theology became prominent again during the Middle Ages, first with Islamic philosophical theologians such as al-Farabi, Ibn Sīnā (Avicenna), and Ibn Rushd (Averroes), and somewhat later with Christian authors including Anselm, Duns Scotus, Thomas Aquinas, and William of Ockham. Many of these authors were influenced by the natural philosophy of Aristotle, which became known through Arabic translations. For instance, Aristotle's concept of a prime mover, a first cause, became important in the development of cosmological arguments in medieval Christianity and Islam. This period saw the development of a variety of sophisticated natural theological arguments.

In the seventeenth and eighteenth centuries, the emergence of the natural sciences led to a thorough reconsideration of the project of natural theology. Devastating religious wars throughout the sixteenth and early seventeenth centuries brought about disenchantment with ecclesiastic authority and organized religion. Since natural theology is concerned with empirical observation and reason, its methodology overlaps to some extent with that of the natural sciences. It became clear that the sciences contradicted scripture, for instance, in the calculation of the age of the earth, its cosmology (heliocentric versus geocentric), and in its model of the origin of species. These scientific challenges to scriptural accuracy put revealed theology under pressure.

Another source of tension for revealed theology was historical biblical criticism, which developed and flourished in eighteenth-century Enlightenment Europe. For the first time in Christian history, the Bible was not seen as the inerrant word of God. Already during the patristic period and beyond, some authors (e.g., Augustine [416] 2002) favored a metaphorical interpretation of some biblical passages that seemed inconsistent with factual knowledge of the world. However, only in the early modern period did a historical view arise of the Bible as the result of a gradual, historical genesis with multiple authors. Biblical scholars began to investigate their historicity through textual analysis (e.g., the internal contradictions and inconsistencies between biblical texts) and by examining the historical and cultural contexts in which these books were recorded.

In this intellectual climate, natural theology became an attractive enterprise. Unlike other forms of theology, it could be conducted independently from scriptural sources and from ecclesiastic authority. But there were also positive reasons for its popularity in the early modern world, particularly the emerging mechanistic worldview, which made analogies between divine and human craftsmanship natural to draw. In England, natural theology flourished between the end of the seventeenth and the middle of the eighteenth centuries. Many natural theologians were natural philosophers (the closest term then for what we now call scientists), such as Robert Boyle and John Ray (see McGrath 2011 for a detailed overview). In the Netherlands, too, natural philosophers such as Bernard Nieuwentijt and Antonie van Leeuwenhoek were impressed by the orderliness and apparent design of the natural world, offering explicit arguments from design.

Many authors have assumed that the early modern project of natural theology collapsed as a result of scientific discoveries, in particular, the theory of natural selection, which provided a viable explanation for design without invoking a designer. Steve Stewart-Williams (2010, 62, n. 16) echoes this familiar sentiment when he writes, "The view that faith rather than reason is the path to true knowledge of God was held long before Darwin promulgated his theory. But the theory provided all the more reason to jettison reason and evidence, and to base one's beliefs entirely on faith (i.e., on nothing)."

There is little historical evidence for this purported role of Darwinism in the destruction of reasoned religious belief. Natural theology continued until well into the early twentieth century. Many theistic authors, such as Asa Gray and Frederick Tennant, reacted favorably to Darwin's theory and sought to incorporate it within their natural theology (see Bowler 2007 for a historical overview). Moreover, philosophical criticisms by Kant and Hume predate Darwin by nearly a century. The design argument, the flagship argument of early modern natural theology, came under attack from naturalistic theories such as the pre-Darwinian transmutation theories that were developed in France, Germany, and the United Kingdom. For example, Jean-Baptiste Lamarck (1809) proposed that transmutation of species occurred through the transmission of acquired characteristics. These ideas enjoyed wide currency across Europe, where they were translated within a few years of publication in several European languages and were further developed by authors in Italy and Belgium. When William Paley wrote his *Natural Theology*, this work already reflected an ongoing debate on the viability of the natural theological project (see chapter 4). It was published at the tail end, not even in the thick of the debate.

A more plausible reason for the decline of natural theology in the late nineteenth century was the incipient influence of methodological naturalism in the sciences. While seventeenth-century natural philosophers still appealed to divine intervention, nineteenth-century scientists adhered more and more to a strict methodological naturalism. This growing popularity of methodological naturalism was partly due to the fact that it worked well. For instance, while Newton still required God to intervene in the movement of heavenly bodies, improved calculations made the appeal to God redundant. However, methodological naturalism was also successful because it was promoted by powerful lobbyists such as Thomas Henry Huxley and other members of the X-Club, a dinner club that aimed, among other things, to excise all appeals to supernaturalism from scientific discourse. In this way, they successfully lessened the competition from amateur-clergymen scientists for posts in scientific institutions and societies (Garwood 2008). This intellectual climate was not congenial to natural theology.

The situation for natural theology became even bleaker by the early twentieth century, when the epistemology underlying science stressed testability and verifiability as essential criteria for empirical claims. This not only widened the gap between natural theology and science but also cast serious doubts on the intellectual legitimacy of natural theology. Since its claims are not verifiable and not empirically testable, according to the logical positivists of the early twentieth century they were either meaningless or false. Logical positivism and its verification principle collapsed in the second half of the twentieth century; the verification principle failed to describe actual scientific practice and could not be coherently applied to itself—the claim that every claim has to be verifiable is itself unverifiable. The newly emerging epistemology was friendlier to natural theology and to philosophy of religion in general. These fields are currently experiencing a revival, with intensive discussion on topics such as the relationship between theism and evil, the nature of divine attributes, and whether or not God knows what people will freely do. Most prominent contemporary practitioners of natural theology, such as Richard Swinburne, Eleonore Stump, and William Lane Craig, are not theologians but philosophers.

Natural theology is conducted in dialogue with naturalistic, philosophical, and scientific worldviews. When religious beliefs are universal and the existence of God (or the gods) is considered obvious, there is little incentive to argue for his existence. By contrast, naturalistic worldviews and ways of thinking prompt the need for explicit and coherent arguments for or against God's existence. Another reason for the close interrelatedness between the sciences and natural theology is an overlap in their basic ways of inquiry: both reason from observation of the empirical world. As we will see in chapters 4, 5, and 8, natural theologians frequently appeal to science, for instance, about the goal-directedness in biological traits (design argument), the big bang theory (cosmological argument), and the concept of scientific laws (argument from miracles). As this book shows, the results of cognitive science of religion are a novel source of inspiration for natural theology. Natural theology, unlike most other forms of theology, does not explicitly presuppose the existence of God. Ideally, natural theological arguments should be intelligible regardless of one's metaphysical assumptions by appealing to observations and intuitions shared by all. In order to be successful, the intuitions to which natural theological arguments appeal should thus be widely shared or at least prima facie plausible. In this book, we will examine what makes these intuitions plausible by considering stable features of human cognition.

What Is Cognitive Science of Religion?

Religious beliefs and practices have been, and continue to be, widespread. About 85 percent to 90 percent of the world's population believes in one or more gods (Zuckerman 2007). Most people worldwide also engage in religious practices at some point in their lives, such as attending weddings, funerals, or visiting religious devotional spaces. Anthropologists regard religion as a universal element of human culture, at least since Edward Burnett Tylor (1871) wrote about Australian aboriginal myths and rituals. He defined religion as beliefs and practices dealing with ancestors, spirits, and other supernatural agents. In the archeological record, evidence for religious belief dates back to at least 40,000 years ago, with imagery of imaginary beings such as sculptures that represent therianthropic (e.g., half-human, half-lion) beings. More tentatively, deliberate burials by *Homo sapiens* and *Homo neanderthalensis* may point to belief in an afterlife, which would push back the first evidence for religious belief to about 110,000 years ago. But these early burials do not contain uncontested grave gifts or other signs that warrant inference to such beliefs. From about 27,000 to 26,000 years ago, we see more definite evidence of intentional burial, including grave gifts in the double-child burial at Sungir, Russia (Kuzmin et al. 2004).

Although there is no generally accepted definition for religion, a suitable starting point is its etymology, the Latin *religare*, to bind or unite. In this view, religions are practices and beliefs that bind communities of people and that link them to a supernatural realm. By this broad anthropological working definition, religion is universal in the same sense that, for instance, music and fire-making are human universals.[4] That is to say, although not all humans have religious beliefs, features of human behavior that are regarded as religious (e.g., belief in supernatural beings, engaging in rituals) are present in all human cultures, and almost everyone has some knowledge of one or more particular religious beliefs and practices, such as the dates and meanings of religious festivals and the properties of supernatural beings.

In the past, the cross-cultural prevalence of religious beliefs formed the main premise of the *consensus gentium* (common consent) argument, which was espoused by authors such as Cicero, John Calvin, and Pierre Gassendi. Roughly speaking, the argument holds that widespread theistic belief constitutes strong evidence for God's existence. The rise of secularism combined with an increased emphasis on individual critical thinking in philosophy and science has led to the demise of this argument (but see Kelly 2011). Nevertheless, the ubiquity of religious beliefs even in the face of well-developed naturalistic worldviews as, in particular, offered by the sciences remains a fact that requires explanation. As the sciences are committed to methodological naturalism, scientists cannot invoke—as the common consent argument did—God's existence to explain the prevalence of religious beliefs. Thus, scientific approaches to religion do not assume God's existence but instead look at natural features of human cognition and social organization to explain its prevalence.

The near-universality of religion has been a focus of intense scientific study. Anthropologists and sociologists traditionally explained religious beliefs and practices as purely cultural phenomena that help humans to make sense of the world and to organize their society (e.g., Durkheim 1915). This cultural layer was regarded as a superstructure that was completely independent from our evolved, biological nature. This treatment of culture as a separate entity prevailed in the social sciences until well into the second half of the twentieth century. It came under increasing pressure as social scientists began to apply insights from evolutionary biology to human behavior (see Pinker 2002 for a popular account). There was a growing recognition of biological constraints, in particular, of the influence of our evolutionary history on our cognitive processes. As a result, cognitive scientists were able to recognize and investigate universal features of religious beliefs and practices across cultures. The cognitive science of religion (CSR) is one of the offshoots of this development. CSR is the scientific study of religion as a natural, evolved product of human thinking.

Cognitive science aims to understand the nature of the human mind and other minds (animals, machines) with a particular focus on how they process information. Information-processing capacities are collectively referred to as *cognition*; they include perception, reasoning, categorization, and memory retrieval. Cognitive science combines insights from many disciplines, including the study of animal cognition, developmental psychology (the study of cognitive development, typically by looking at cognitive processes in infants and children), neuroscience, cultural anthropology, artificial intelligence, and philosophy. Cognitive science began in the 1940s and 1950s as a study of intelligent, adaptive behavior in humans and non-human animals, with the aim of reproducing such behavior in artificial systems. Since the 1980s, cognitive scientists have studied how adults, children, and animals learn about the world and how they can master complex skills such as speaking their native language, counting, and using tools and other artifacts. More recently still, cognitive scientists have also turned their attention to the study of culture. They want to find out why some beliefs are culturally widespread while others are rare, and how culturally transmitted beliefs change over time. Religious beliefs and practices fall within this burgeoning research interest.

Although cognitive science is a young discipline, it has ancient roots in natural philosophy. Questions on the workings of the mind hold a prominent place in natural philosophy at least since antiquity. For instance, in the *Meno*, Plato ([ca. 380 BCE] 2000) wondered how it is possible for humans to acquire knowledge of abstract properties such as morality and geometry. The questions that natural philosophers attempted to answer are very similar to those posed by cognitive science today. In the course of the sixteenth through nineteenth centuries, natural philosophy gradually transformed into diverse branches of modern science, with its emphasis on empirical testability and experiment. Nevertheless, the sciences still maintain an important aspect of natural philosophy: they aim at a better understanding of the natural world. Scientists not only want to have theories that help them achieve practical goals (such as building machines and developing effective medicine) but that also help them better understand themselves and the world they live in. CSR is in line with this natural philosophical aspect of science. Titles of books such as *Why Would Anyone Believe in God?* (J. L. Barrett 2004), *Religion Explained* (Boyer 2002), and *Supernatural Agents: Why We Believe in Souls, Gods, and Buddhas* (Pyysiäinen 2009) testify to this natural philosophical concern: these authors want to understand why religious beliefs and practices are widespread across time and cultures.

William James's *The Varieties of Religious Experience* (1902) is a precursor of CSR. Like many of his contemporaries, James assumed that religious experiences (especially mystical experiences) were central to explaining religion. Relying on medical knowledge available at the time as well as historical sources on mysticism, he postulated that religious experiences are caused by medical conditions such as epileptic seizures. CSR no longer considers such rather exceptional medical conditions as the main causes of religious belief. As we will see in the next chapter, most cognitive scientists of religion hold that religious beliefs and practices find their roots in normal, everyday cognitive capacities that are present in all neurotypical humans and that arise early and spontaneously in young children.

Jean Piaget is another important precursor of CSR. He recognized that young children attribute superhuman, godlike properties such as omniscience and omnipotence to their parents. Sigmund Freud (1927) also thought that belief in supernatural beings is a projection of parental qualities onto an imagined being. However, Piaget (1929) surmised that children's projection of parental qualities on God was driven by an intellectual need to understand the world, whereas Freud claimed that this projection was the result of an emotional need for attachment. These views are still being considered today, albeit in updated forms (see e.g., Kelemen 2004; Rossano 2010).

CSR scholars examine behaviors—such as prayer and ritual—and beliefs in, among others, supernatural agents, an afterlife, and body–soul dualism. Although CSR is not a well-delineated research program, its practitioners share several underlying unifying assumptions that are characteristic for cognitive science as a discipline. We will look at some of these assumptions in more detail in the next chapter and briefly summarize them here:

• Religious beliefs and practices result from normally functioning human cognitive processes. From a cognitive point of view, there is nothing exceptional about religion. CSR examines religious beliefs in terms of ordinary beliefs about objects and agents, and religious behavior in terms of normal human behaviors.

• Religion is not a purely cultural phenomenon. The architecture of the evolved human mind constrains and guides the development of religious beliefs. This, however, does not preclude an important role of culture.

• The human mind is composed of several domain-specific cognitive capacities that underlie, among other things, our ability to acquire particular religious beliefs.

The subject matter of CSR is diverse, reflecting the variety of religious beliefs and practices it studies. One of its main topics is the cross-cultural prevalence of belief in supernatural agents such as gods, ghosts, and ancestors. CSR scholars are also intrigued by the nonnatural properties humans attribute to themselves, such as the existence of a soul, spirit possession, and the afterlife. They also investigate religious practices. Rituals, the most conspicuous of these, are intriguing because of their costliness in terms of time, energy, and precious offerings. Harvey Whitehouse (2004) has noticed a distinction between two types of rituals: some are performed with high frequency, streamlined and highly codified, and often tied to religious doctrines or myths. Others are rare, involving intense experiences of pain and deprivation, for example, traditional rites of passage in sub-Saharan African and Papuan cultures. Other religious practices under examination include prayer and various tokens of religious membership such as dress codes and observance of dietary restrictions. Since religion is not an isolated phenomenon but intertwined with other dimensions of the human condition, CSR also investigates how religion interacts with other aspects of cognition and culture. One of the most intensive areas of research in this field is the relationship between religion and morality, which we will consider in more detail in chapter 6.

For most of these areas of inquiry, CSR research is germinal, and at present there is an imbalance between theory and practice. Many CSR theories have not been empirically tested. Some religious beliefs and behaviors have hardly been studied. For example, although prayer is a central element of religiosity across cultures, to date only a handful of studies have looked at its cognitive underpinnings (e.g., Barrett 2001; Boudry and De Smedt 2011; Schjoedt et al. 2009). The relative scarcity of empirical work makes it sometimes difficult to assess the merits of these hypotheses. Not only is its range still constrained, there is also a clear need for cross-cultural replications. Most CSR research is still conducted with Western, middle-class participants (children and adults), whose responses may not be representative of the world's population at large—they are WEIRD (from Western, Educated, Industrialized, Rich, and Democratic societies), to use the contrived acronym of Henrich, Heine, and Norenzayan (2010). While there is some research with non-Western populations such as the Shuar from Ecuador and the Vezo from Madagascar, a more widespread cross-cultural replication of CSR studies is needed. These are limitations that should be borne in mind when relying on theories and evidence from CSR.

Summary

This chapter has outlined two central questions about religion: what is its foundation in reason (natural theology) and what constitutes its origins in human nature (CSR)? While these questions are often considered independently from each other, there are interrelations between them that deserve further scrutiny. First, cognitive science has the potential to uncover the enduring appeal of natural theology and the intuitive plausibility of its arguments. By looking at the evolved structure of the human mind, we can learn something about the origin of intuitions that underlie natural theology. Second, CSR has the potential to challenge the reasonableness of religious beliefs.

This brief survey of the history of natural theology indicates that natural theology enjoys an enduring appeal. Natural theologians connect our reason and experience of the world to the question of God's existence and attributes. They do so in dialogue with the empirical sciences. CSR is a fairly recent research program with natural philosophical roots that aims to understand the causal origins of religious beliefs. In the remainder of the book, we will examine the implications of CSR for natural theology.

2 The Naturalness of Religious Beliefs

Most CSR scholars propose that religion is universal because it is natural, the product of normally functioning human cognitive processes. What does this claim mean? How can we define what is natural and what is not, and how can we gauge whether normal human cognition spontaneously produces religious beliefs? To answer these questions, this chapter provides a detailed review of the theoretical assumptions about the human mind that underlie CSR. We will investigate what CSR scholars and cognitive scientists in general mean when they say that religion is natural, and how they regard the relationship between psychological dispositions and cultural religious practices.

The Human Cognitive Toolbox

Cognitive scientists from several domains, including cognitive neuroscience, evolutionary psychology, and developmental psychology, hold that the human brain is not a blank slate but that it is composed of several cognitive specializations. Humans are equipped with an evolved cognitive toolbox containing specialized mental tools that help them deal with distinct cognitive tasks such as finding their way, interacting with other people, recognizing what they can eat, and predicting the movements of inanimate objects. This hypothesis of a diverse and specialized cognitive toolkit is a central assumption in CSR and thus requires closer investigation. We will here discuss neuroscientific, evolutionary, and developmental psychological theories of cognitive specialization and their relevance for CSR.

The idea that human cognition is composed of specialized processing units can be traced back to the eighteenth-century discipline phrenology, which consisted of correlating bumps on the skull with cognitive faculties such as love of one's offspring, semantic memory, and the gift of music. Although this research was later debunked, phrenology was revolutionary in its identification of the brain as the locus of human cognition. Prior to this research program, physicians assumed that the nerves were hollow, containing a fluid that was the substance of the soul; the main function of the brain was its filtration and purification (Greenblatt 1995). In the nineteenth century, physicians such as Paul Broca and Carl Wernicke discovered that brain lesions (local forms of brain damage) could lead to selective speech impairment, in this way reliably identifying parts of the brain that were involved in specific cognitive functions. This lesion method is still used in cognitive neuroscience today.

A more direct way to study which regions of the brain are involved in performing specific tasks is provided by functional neuroimaging techniques, which were developed in the later decades of the twentieth century. All neuroimaging techniques exploit the fact that brain regions that are more active require more energy (glucose) and oxygen than other areas. These techniques measure differential brain activation after presentation of a relevant stimulus, and compare these activations to a carefully chosen control stimulus. If this effect is constant across subjects and if it is reproducible, the cerebral parts that are more active after presentation of the test stimulus compared to a control condition are taken as neural correlates for the task that the stimulus probes. This has led to the discovery of many specialized neural regions. Typically, a given neural region is involved in more than one task. For example, Broca's area, which was originally implicated in grammar comprehension, is also involved in imitation and in understanding the structure of musical compositions. Also, cognitive tasks usually involve the simultaneous recruitment of several neural regions. For example, the ability to infer mental states is described as a network of anatomically defined regions, each specialized in a specific subtask such as intentionality detection (superior temporal sulcus), explicit representation of states of the self (medial prefrontal cortex), and inferring emotions from facial expressions (anterior region of the superior temporal gyrus).

Functional magnetic resonance imaging (fMRI) is currently the most popular neuroimaging technique. It captures the oxygenation of the blood flow within the brain of a participant who is lying in a scanner. Its basic assumption is that areas of the brain involved in a task need more oxygen and that oxygenated blood has different magnetic properties compared to non-oxygenated blood. The brain during (wakeful) rest is taken as a baseline. Techniques such as fMRI have been applied to the study of religious cognition, but their use is currently quite limited. For one thing, the participant is in a hospital setting, required to lie still within a large, noisy machine, not an ideal environment to engage in religious activities like prayer or meditation. Moreover, it rules out analysis of ritual actions that require bodily movement, such as temple dancing or prostration. Until recently, neuroscientists were especially interested in understanding exceptional and expert religious practices, such as meditation by Buddhist monks and mystical experiences of Franciscan nuns (Azari et al. 2001; Newberg et al. 2001; Beauregard and Paquette 2006). As a result, the neuroscience of religious belief and practice remains a small field. However, an increasing (yet still modest) number of studies with more mundane religious practitioners tentatively suggests that religious beliefs and practices rely on normal cognitive functions. For example, when subjects are considering God's emotions, such as his anger or emotional involvement, this activates brain areas implicated in theory of mind (Kapogiannis et al. 2009). Similarly, when engaged in spontaneous prayer, brain areas involved in social interactions with other people, such as the temporopolar region, the medial prefrontal cortex, and the temporoparietal junction, are activated (Schjoedt et al. 2009).

A direct study of brain activity is not the only way to find out how human cognition is structured. Evolutionary psychology and human behavioral biology study cognition by probing the selective pressures that have shaped cognitive capacities. Animals behave adaptively and flexibly under a wide range of circumstances: they find food, flee from predators, care for offspring, and interact with group members. Evolutionary approaches to cognition explain this efficiency by proposing the existence of cognitive modules, cognitive systems that are functionally specialized to deal with specific domains of knowledge about the environment, such as finding food (e.g., recognizing plants) and interacting with others. Each of these systems has a specific task that is relevant from the point of view of natural and sexual selection: they contribute directly or indirectly to the fitness of the organism (Cosmides and Tooby 1994b).

Why assume that animals have several functionally specialized systems rather than a single, cognitive mechanism for dealing with adaptive problems? Many adaptive problems require functionally incompatible solutions. Humans, for example, can predict the motions of animate and inanimate objects. These tasks probably demand functionally incompatible cognitive tools: predicting the movement of an inanimate object requires the ability to understand its velocity, trajectory, and momentum, whereas evaluating the motion of an agent additionally requires attribution of goals and intentions. Even infants expect that inanimate objects behave differently from agents (Kuhlmeier, Bloom, and Wynn 2004). Indeed, it turns out that the detection of inanimate versus animate motion activates distinct areas in the human brain (Martin and Weisberg 2003). Given that functional incompatibility has rarely been explicitly researched as a property of natural cognitive systems, it remains as yet unclear to what extent our cognitive architecture is filled with functionally incompatible structures.

Cognitive capacities are often subdivided in two categories: adaptations, which deal with specific adaptive problems, and byproducts, which do not fulfill adaptive functions but emerge as byproducts from adaptations. Applied to religion, *adaptationist* explanations of religion propose that religious beliefs and practices serve a direct adaptive function, in particular, that they enhance human cooperation (see Schloss and Murray 2011 for a review). Accordingly, religion has played and still plays a functional role in increasing the fitness of religious individuals in a range of environments. By contrast, *byproduct* explanations of religion propose that religious beliefs are not directly adaptive but that they result from the normal workings of human cognitive systems (Bloom 2007). For example, religion may be a byproduct of our tendency to discern agents in the environment, it may be a product of our tendency to see teleology and design in the natural world, or it may arise from the intuitive distinction we draw between minds and physical objects.

Next to adaptationism and byproduct views, there is a third, more recent way to look at cognitive specializations. According to this view, cognitive adaptations are flexibly redeployed in the face of novel cultural challenges. The classical adaptationist and byproduct views are quite Cartesian in that they assume a strict division between the mind and the world. By contrast, the *redeployment view* acknowledges that the cultural environment plays a role in the development of the brain and can shape the functions of evolved cognitive mechanisms to deal with new tasks, such as driving or playing chess.[1] As embodied beings, our cognitive system is responsive to the specific demands that the cultural environment poses. For example, reading and writing are skills that emerged less than 10,000 years ago, too recent for natural selection to shape our brains for literacy. According to Dehaene and Cohen (2007), brain areas involved in language comprehension and the visual discrimination of small details are co-opted in these tasks. The visual word form area (VWFA) is a highly specialized and localized brain area consistently involved in reading. Its original function is to recognize objects in the environment and to discriminate between them. Neurons in the homolog of the VWFA in monkey brains fire selectively when a particular object is presented, regardless of viewpoint; for example, a spoon will be recognized as such, regardless of whether it is shown from the top or the side. This makes it very suitable to respond selectively to a specific grapheme, such as the letter A, as it can generalize across font, uppercase, and lowercase; neurons in the VWFA can be trained to recognize "a" and "$\mathcal{A}$" as the same letter. Religion is a relatively recent phenomenon (110,000 or perhaps only 40,000 years old, according to archeological finds), so we can expect that this cultural domain co-opts older cognitive capacities. As we have seen, the brain handles interactions with God in a way that is similar to interactions with other persons. Throughout this book, we will see how the even more recent domain of natural theology also redeploys preexisting cognitive specializations such as the abilities to detect design and to infer causes.

Concurrent with the neuroscientific and evolutionary approaches, developmental psychologists have developed a theoretical framework to explain features of human cognition. Of these three approaches, theirs has been the most influential in CSR. Developmental psychologists study the cognitive capacities of infants and young children, and examine changes in cognition as they mature. Authors such as Elizabeth Spelke and Susan Carey propose that children can acquire knowledge about the world through domains of *core knowledge*. These are innate bodies of knowledge and inference mechanisms that deal with distinctive categories of objects in the world, including inanimate objects (intuitive physics), conspecifics (intuitive psychology), animals and plants (intuitive biology), and geometry and number (intuitive mathematics) (Spelke and Kinzler 2007).

Core knowledge undergoes predictable developmental changes over time. For example, in the domain of intuitive psychology, children under four typically cannot take into account false beliefs in explicit reasoning tasks, whereas older children can. Core knowledge can also interact with the cultural environment, giving rise to fascinating mental models. For example, most children experience the ground as flat, which leads to the intuition of a flat Earth. They also have the experience and intuition that unsupported objects fall downward, which provides further support for a flat Earth model. To harmonize this with scientific information they gradually acquire (such as satellite images of the Earth), they reconstruct the Earth as a flat, circular object. While both Indian and American first-graders make disk-shaped models of the Earth, only the former will state that this disk is surrounded by water, in accordance with Indian folk cosmology (Samarapungavan, Vosniadou, and Brewer 1996). Interestingly, Australian preschoolers, possibly because of their historical ties to the Old World and the latter's conceptualization of Australia as being at the other side of the planet, are much more advanced in their understanding of the shape of the Earth compared to British children of the same age (Siegal, Butterworth, and Newcombe 2004).

In spite of developmental changes and sensitivity to culture, most proponents of core knowledge argue that these domains are innate and domain-specific, and that they continue to play a role in inferences adults make in their everyday life, even in their scientific practice (Carey and Spelke 1996). How can one decide whether a body of knowledge is innate? Although innate traits can arise late in development (e.g., secondary sexual characteristics), developmental psychologists typically take the early emergence of a given domain of knowledge as strong evidence for its innateness. They rely on a version of the poverty of the stimulus argument that Samuels (2002) has termed "the argument from early development." A given domain of knowledge is innate if it arises so early in development that it could not have been learned through experience. For example, infants at five months old are able to predict the results of simple additions such as $1 + 1 = 2$ and not 1, at an age at which they do not have the motor control to test whether this result always holds (Wynn 1992).

Because infants cannot give verbal responses, much research in developmental psychology is based on the *violation of expectation procedure*.[2] This technique exploits the propensity of humans and other animals to look longer at unexpected than at expected events. Our knowledge of the world enables us to make predictions of how objects will behave. For example, we expect an object to fall downward and not upward. An upward falling object under normal gravity conditions would cause surprise and longer looking times. *Surprise* is used as a shorthand descriptor to denote a state of heightened attention or interest caused by an expectation violation. Prior to the test trials, infants are exposed to habituation or familiarization trials to acquaint them with various aspects of the test events. With appropriate controls, evidence that infants look reliably longer at the unexpected than at the expected event is taken to indicate that they possess the expectation under investigation, detect the violation in the unexpected event, and are surprised by this violation.

Since many authors in CSR are developmental psychologists by training or through their research (e.g., Justin Barrett, Deborah Kelemen, Paul Bloom), it is not surprising that core knowledge has played an important role in the theoretical developments in current CSR scholarship. Religious beliefs are assumed to be a byproduct of the normal workings of core knowledge domains. According to Paul Bloom (2007, 150), "Certain early-emerging cognitive biases ... give rise to religious belief. These include body–soul dualism and a hyper-sensitivity to signs of agency and design. These biases make it natural to believe in Gods and spirits, in an afterlife, and in the divine creation of the universe. These are the seeds from which religion grows."

Within this broad framework, there is disagreement among developmental psychologists about whether religious beliefs are the spontaneous and unlearned outputs of core knowledge, as Bloom, Kelemen, and Barrett assume, or whether they require substantial cultural input for their development, as Rita Astuti and Paul Harris believe. For example, developmental psychologists debate the extent to which belief in an afterlife is culturally mediated. Some (Bering, McLeod, and Shackelford 2005; Bloom 2004) believe that this emerges spontaneously in young children prior to, and independent from, instruction and education. By contrast, others (Harris and Giménez 2005; Astuti and Harris 2008) have found that children's assessment about whether or not a person lives on after death depends on circumstances, in particular on narrative context: they make more continuity judgments (i.e., attribute continued properties like emotions, dislikes, and bodily functions to the deceased) when the story plays in a religious setting than when it plays in a hospital. Harris and coauthors take this as evidence that culture plays a decisive role in the development of beliefs about the afterlife. In line with the redeployment framework, we would expect that core knowledge domains are indeed flexibly redeployed in the specific religious systems people grow up in.

Intuitive Ontologies as Natural Modes of Reasoning

In the previous section, we surveyed concepts of cognitive specialization from diverse disciplines. For the purposes of this book, we will use an eclectic notion of human cognitive specialization termed *intuitive ontologies*.[3] This view reflects how many cognitive scientists of religion conceptualize human cognition (e.g., Boyer 2002; J. L. Barrett 2004; Pyysiäinen 2009). Intuitive ontologies are an evolved set of category-based expectations that emerge early in development and that guide reasoning about physical, psychological, and biological phenomena.[4]

The concept of intuitive ontologies combines elements from cognitive neuroscience, evolutionary psychology, and developmental psychology. It draws on neuroscience in the view that intuitive ontologies are subserved by a stable neural architecture; for example, knowledge of living things and knowledge of artifacts are represented in different areas of the brain (Caramazza and Mahon 2003). It borrows the idea of functional specialization from evolutionary psychology in that intuitive ontologies are associated with computationally complex survival problems, including finding food, avoiding predators, and handling tools. It also incorporates core knowledge in that intuitive ontologies are relatively broad and emerge early in development without instruction. They are termed "intuitive" because they are not products of deliberate reflection or scientific investigation. Intuitive ontologies that will return in the remainder of the book include intuitive physics (how inanimate objects move), intuitive psychology (what agents think and how and why they act), intuitive biology (how livings things develop and behave), and intuitive engineering (how artifacts are made and function).

Intuitive Physics

Intuitive physics helps us understand and predict the motions and behaviors of inanimate objects. Looking time experiments (e.g., Spelke, Phillips, and Woodward 1995) indicate that infants as young as four months have expectations about the behavior of inanimate objects. They are surprised when a solid-looking object suddenly falls apart without any apparent external cause or when an inanimate object apparently causes the motion of another one without direct contact. Elizabeth Spelke (1990) proposes that three principles regulate infants' cognition about objects: continuity (objects that are out of sight continue to exist), contact (inanimate objects must be in direct contact to influence each other's behavior; there is no action at a distance), and cohesion (inanimate objects do not fall apart without external cause).

We share some principles of intuitive physics with other animals. For example, the primatologist Daniel Povinelli (2000) found that chimpanzees also understand the principles of contact and continuity. However, humans may be unique in their ability to understand physical phenomena in terms of invisible generative causes such as gravity and momentum (see chapter 5). The intuitive belief in invisible forces such as gravity underlies the prediction of how physical objects will behave. These intuitive predictions do not always match actual physical events: for example, most people make predictions about physical events in line with medieval impetus theory,[5] say, by erroneously expecting that a ball launched by a sling will continue its course in a spiral trajectory (McCloskey, Caramazza, and Green 1980); most also wrongly believe that an object dropped by a running person will fall straight down to the ground (McCloskey, Washburn, and Felch 1983). These errors are not due to a lack of schooling, since the subjects who made these mistakes had been taught Newtonian physics at least at the high-school level and some even at the undergraduate level. Moreover, under conditions of rapid judgment, where participants have to make very quick predictions regarding the trajectory of inanimate objects, even professional physicists give responses that are more in accordance with intuitive physics than with Newtonian mechanics (Kozhevnikov and Hegarty 2001).

Intuitive Psychology

Humans have an intuitive psychological theory (also termed theory of mind) that explains actions by attributing internal, unobservable mental states, such as beliefs, desires, and intentions. It develops during the first year of life, culminating in the ability to verbally solve false belief tasks between four and five years of age. By that age, children realize that the mental representation of a situation may be different from the situation itself (Frith and Frith 1999). Their understanding of false beliefs is gauged by tasks that involve deception, such as the so-called Sally-Anne task, during which the child has to predict where an agent (Sally) will look for a previously hidden object that was moved to another location by a different agent (Anne) during Sally's absence. Solving this false belief task requires the attribution of a false belief to Sally. It is a cross-culturally robust observation (Callaghan et al. 2005) that children routinely fail such verbal false belief tasks at age three but can solve them by age five.

Recent experimental studies indicate that an implicit understanding of false beliefs appears earlier in life. Although most three-year-olds fail verbal false belief tasks, infants succeed at implicit, nonverbal versions of these tasks. For example, fifteen-month-olds look longer when an agent searches for an object in a container where it was moved to in the agent's absence and expect her to look for the object in the location where she originally hid it (Onishi and Baillargeon 2005). Another looking-time experiment (Kovács, Téglás, and Endress 2010) pushes this further back to seven months of age. The latter experiment indicates that infants are influenced by the inferred beliefs of other agents, even if this conflicts with their own beliefs (see also chapter 3). It seems likely that the ability to infer beliefs, especially false ones, is a uniquely human capacity, as nonhuman apes have consistently failed nonverbal false belief tasks (Call and Tomasello 1999).

It is important to note that intuitive psychology (like all intuitive ontologies) consists of several, more fine-grained capacities; it cannot be reduced to an ability to infer beliefs. Our intuitive psychology also includes detecting eye-gaze direction and inferring goals. In many cases of day-to-day interaction, these less computationally demanding actions likely play an important role. For instance, the detection of eye-gaze direction enables us to infer someone's intentional state without having to attribute explicit beliefs to this person. Infants are able to detect from birth whether someone makes eye contact with them (Farroni et al. 2002). The widespread occurrence of eyespots in animal mimicry indicates that this ability is phylogenetically ancient. Next to eye contact, gaze direction can be used to infer the attention of an agent to things other than oneself. Not only apes, but also a variety of domesticated animals (e.g., horses, goats, dogs) are proficient in this (see Itakura 2004 for an overview). The ability to infer goals is present in nine-month-olds, who can predict, for example, which path an agent will take if it wants to reach a specific location (Gergely et al. 1995). A growing number of empirical studies also indicates that chimpanzees can infer goals even if they fall short of representing full belief states (e.g., Hare, Call, and Tomasello 2001).

Intuitive Biology

At around four years of age, children develop an intuitive biology (Ahn et al. 2001) containing rich inferences about inheritance (offspring resembles its parents) and patterns of growth and development (members of the same species typically go through the same irreversible patterns of growth). Neuroscientific evidence (e.g., Caramazza and Shelton 1998) indicates that intuitive biological knowledge is dissociable into knowledge about animals and knowledge about plants: lesion studies show that some patients can retain knowledge of plants while losing the ability to reason about and recognize animals. Nevertheless, inferences about heredity and development are common to both categories.

One of the core intuitions underlying these rich inferences is *psychological essentialism*. This term is used by developmental psychologists to distinguish it from other forms of essentialism, notably philosophical essentialism as developed by Aristotle. It denotes a psychological tendency to attribute essences to biological and natural kinds. Young children and adults hold the belief that living kinds possess an unchangeable hidden essence, which causes their adult form and behavior (Medin and Ortony 1989). For instance, they believe that surgically altering a dog to look like a raccoon does not actually transform the dog into a raccoon (Keil 1989): it is the inside essence that matters, not the outside superficial appearance. Four-year-olds also predict that an apple seed planted in a pear orchard will still grow into an apple tree, or that a young kangaroo raised by goat foster parents propels itself by hopping rather than climbing even if it never sees another kangaroo in its life (Gelman and Wellman 1991).

Teleology, the tendency to reason about the parts of living things (and even about living things in their entirety) in terms of purposes, is another important component of intuitive biology. It enables children to learn useful facts about organisms. Three-year-olds can infer the diet of an unfamiliar animal based on the shape of its beak and can predict that an unknown mammal with webbed feet has an aquatic lifestyle (Kelemen et al. 2003). Preliminary cross-cultural studies indicate that teleology is not only deeply engrained in Westerners but also appears in other cultures such as the Shuar, Native Americans from the Andes. The Shuar also classify animals and plants according to their functional parts and think about them in terms of goal-directedness (H. C. Barrett 2004).

Intuitive Engineering

A growing number of developmental psychologists (e.g., Kelemen and Carey 2007) are convinced that artifacts constitute a distinct domain of intuitive knowledge with its own set of inductive inferences and assumptions. As we shall see in more detail in chapter 4, humans have an intuitive *design stance*, by which they infer the function and identity of artifacts by the (inferred) intention of the designer. When considering an artifact, they tend to keep the intended identity or use of the object in mind. Humans find the original function of an artifact more important than its current use. For example, adults judge that an object designed for exercising back muscles is still a training instrument even if it is currently exclusively used for stretching clothes (German and Johnson 2002). The design stance develops gradually between the ages of about nine months and six years (Casler and Kelemen 2007). Two-year-olds who are exposed to a novel artifact that is used for a specific purpose will continue to employ it in this way, even if they only saw it demonstrated once, a phenomenon known as fast mapping. Moreover, they will avoid using the artifact for other equally feasible purposes (Casler and Kelemen 2005).

The key signature of a fully developed design stance is *functional fixedness*: children over the age of five become fixed upon the intended design and function of an object and tend to see it as only functioning in a particular way. The Shuar, who have little material culture, are also hindered by functional fixedness (German and Barrett 2005). This is especially intriguing given that they often recycle their artifacts for other purposes. Yet, Shuar participants will insist that an object that was originally made for catching fish is a net, even if it is currently exclusively used as a hammock. Our intuitive ontological assumptions about artifacts may constitute a human-specific cognitive adaptation, possibly evolved as a result of hominids' extensive reliance on tools during at least the last 2.5 million years (Semaw et al. 1997).

Naturalness

CSR authors frequently appeal to ambiguous terms such as *natural* to denote the relationship of religious beliefs and practices to human cognition. The cultural variability of religious beliefs and practices indicates that natural does not mean the opposite of cultural: the cultural richness of religion can perhaps best be illustrated by the fact that anthropologists typically take months or years to learn about the religious beliefs of other cultures. Rather, this term is used in a more technical sense.

Although students of religion have widely claimed that religious beliefs and practices are cognitively natural, they do not use a unified concept of naturalness. Some (e.g., Dennett 2006) have understood this as an onto-logical claim, namely that religion is not supernatural, but entirely a product of natural human cognitive processes and social interactions. To others (e.g., Boyer 2002), naturalness means that religious ideas are easily acquired because they fit well with the structure of human cognition: religious ideas match early developed cognitive biases. A stronger version of the naturalness thesis (e.g., Barrett 2010) holds that humans are predisposed to believe in religious entities: religiosity is a mode of thinking that emerges spontaneously as a result of the interaction between our cognitive processes and the external world. This requires little, if any, cognitive effort. The strongest versions of the naturalness-of-religion thesis are nativist (e.g., Bering 2006, 2011): they assert that religious concepts such as ghosts and punishing deities emerge spontaneously from the innate structure of our minds and that such concepts require only modest, if any, cultural input.

Robert McCauley (2000, 2011) has argued that religion is natural, and science is not. Since his discussion presents the most detailed conceptual analysis of this term to date, we will consider it in more detail. He uses *natural* (or, more specifically, *maturationally natural*) in a fairly restrictive sense, namely as those properties of human cognition that arise early in development, without explicit instruction. Examples include speaking one's first language, walking, and chewing. Other distinguishing features of maturationally natural capacities include their lack of material support (they do not require artifacts to acquire them), the fact that we spontaneously engage in them, and that they often address basic problems humans face (e.g., communication with others, getting from one place to another). McCauley also speaks of *practiced naturalness*, which includes skilled actions, such as cycling and reading, that through diligent practice have become second nature to us. But to McCauley, the sense of maturationally natural is the more fundamental of the two. Arguably, it is the one that CSR has paid most attention to. According to McCauley (2011, chapter 4), acquiring folk religious beliefs is maturationally natural. Although there is some cultural variation, such beliefs easily latch on to early-emerging cognitive capacities. For instance, speaking in tongues (glossolalia) is a phenomenon that appears in several religions (e.g., Pentecostalism, a number of African local religions). While the utterances produced in this way sound like an unknown language, it turns out that glossolalists use only a subset of the phonemes of their native language and restrict themselves to pitch contours, intervals, and stress patterns that are characteristic of their mother tongue. Glossolalia keys in on our auditory sensitivity to language-like sounds and prompts a spontaneous search for meaning—indeed, in many Charismatic Christian communities, this is seen as a manifestation of the Holy Spirit.

McCauley (2011, chapter 3) also outlines *unnatural* domains of human skill and knowledge, such as science. Science imparts concepts and worldviews that are contrary to everyday knowledge and experience. For instance, the heliocentric model of the Earth orbiting the Sun contradicts our visual observation of the Sun moving across the sky, particle physics is at odds with our intuitive understanding of solid objects, and Newtonian physics violates our intuitive predictions of the motions of objects based on their impetus. An intriguing experiment that involved rapid judgments indicates that recalling scientific information requires continuous effort, even for scientific facts we know well (Shtulman and Valcarcel 2012). For example, it takes people less time to judge that "The Moon revolves around the Earth" is true than it takes them to affirm that "The Earth revolves around the Sun" (presumably because we see the Moon moving across the sky, but we do not see the Earth's revolution). Being a professional scientist requires extensive cultural support, a hallmark of cognitive unnaturalness. The skills required for gathering and interpreting evidence and for testing and generating hypotheses demand training and practice. In light of this, it is unsurprising that the vast majority of the world's population does not engage in science and may not fully understand its basic concepts.

However, scientific expertise can allow for practiced naturalness. For instance, amplifying DNA requires skills that are learned in the lab through practice and experience, and an experienced archeologist can learn to visually assess, without using complicated measuring techniques, whether a particular coloration of the soil is of anthropogenic or non-anthropogenic origin (e.g., traces left by wooden fences or by digging earthworms). While we will qualify the term *cognitively natural* or simply *natural* in the next chapter, it remains for now useful shorthand for how cognitive scientists understand this term.

Intuitiveness

Like philosophers, CSR authors frequently use the terms *intuition* and *intuitive*. These terms do not have the same meaning in both disciplines, although there is some semantic overlap. Intuition comes from the Latin *intueri*, which means looking at or considering. In a broad sense, intuitions refer to those thoughts or assessments that come about without (conscious) inference or deliberation. They appear to be true by virtue of a kind of credibility of their own, rather than by reasoning from one or more premises. For instance, the proposition "If you drop this book, it will fall downward" is intuitive because the belief is immediately plausible without appeal to inferential reasoning, such as, "All objects in proximity of the Earth are subject to its gravitational pull; this book is close to the Earth, therefore, if released, it will fall toward the center of the planet with a velocity of 9.81 m/s^2."

Philosophers since René Descartes have noticed similarities between intuitions in philosophy and normal perception. Like perceptual experience, intuitions are immediate. Descartes ([1619] 1985, rule 3, ATX 368), in his *Rules for the Direction of the Mind*, calls intuitions a type of perception. Their most distinguishing feature is their lack of movement from premises to conclusions (by aid of memory) that is characteristic for reasoning. Intuitions can simply be grasped without much, if any, cognitive effort. Despite his radically different philosophical outlook, John Locke too thought that intuition was a non-inferential source of knowledge: it occurs when we spontaneously perceive a connection between several ideas. We can mentally perceive that "*white* is not *black*, that a circle is not a triangle, that *three* are more than *two*, and equal to *one and two*" (Locke 1690, 264). Remarkably, to Locke (1690, 264), intuitions afford a higher degree of certainty than inferential reasoning: "This part of knowledge is irresistible, and, like bright sunshine, forces itself immediately to be perceived, as soon as ever the mind turns its view that way; and leaves no room for hesitation, doubt, or examination, but the mind is presently filled with the clear light of it. 'Tis on this intuition, that depends all the certainty and evidence of all our knowledge."

However, in the next centuries, the picture of intuition as a source of indubitable knowledge changed profoundly. The success of the experimental method in the sciences led to a distrust of formulating beliefs without doing empirical observations; for example, it seems intuitively plausible that heavier objects fall faster than light ones, but it takes empirical observation to correct for this. Also, within philosophy, authors such as Laurence BonJour (1985) have argued that intuitions can never be a source of knowledge because they do not depend on inferential justification, that is, on reasons for why the intuition is thought to be true. Nevertheless, because philosophy is still for the most part an armchair discipline, most of its practitioners (e.g., van Woudenberg 2005) reject this skeptical conclusion and maintain that intuitions can be, and often are, sources of knowledge.

In cognitive science, the term intuition arises in dual-processing accounts of reasoning, which are becoming increasingly influential. They distinguish between two types of thinking, fast and slow (Evans 2008), which give rise to intuitive and reflective beliefs respectively (Sperber 1997). Intuitive beliefs emerge spontaneously through an interaction of our cognitive architecture with the external world. They are fast and effortless, requiring no deliberation or conscious reflection. Consider the following propositions: "This desk at which I am writing is a solid, wooden object" and "This desk consists mostly of empty space, dotted with tiny particles that are too small for anyone to see with the naked eye and that are held together by the strong nuclear force." Many Western-educated adults hold both beliefs, although they seem incompatible. The first belief is intuitive. It can arise through normal perception (e.g., seeing the desk) or through testimony (e.g., even if one has never seen a Regency desk, one can form an idea about what they are like through the accounts of others).

Reflective beliefs, on the other hand, are formed through slow, deliberate, and conscious reasoning processes. Many cognitive scientists argue that our capacity for reflective reasoning has an evolutionary origin. For example, Dan Sperber (1997) hypothesizes that this capacity finds its origin in intuitive psychology. Reflective beliefs allow us to assess the claim "Peter believes that Sydney is the capital of Australia" as true, even though the proposition "Sydney is the capital of Australia" is false. This helps us to understand that others have false beliefs and to correctly bracket off false beliefs from our own correct beliefs ("Canberra is the capital of Australia"). Peter Carruthers (2006) links the emergence of reflective reasoning to language: thanks to our linguistic skills, we are able to evaluate propositions that are not intuitively obvious or hard to understand, such as the scientific testimony that underlies our reflective belief that desks actually consist of empty space dotted with particles.

Although many intuitions that cognitive scientists have identified are maturationally natural (e.g., intuitive ontologies), intuitions can also be shaped through practiced natural skills. For instance, mathematicians experience provability intuitions, strong, unreflective beliefs about whether a mathematical conjecture is provable. The Goldbach conjecture (every even number greater than 2 can be expressed as the sum of two primes) has not yet been proven (and will perhaps never be), but many mathematicians have the intuition that the statement is true and in principle provable. Mathematicians also rely on intuitions to judge whether a proof is sound or not before going through the actual work of checking the details (Thurston 2006). Plumbers have an intuitive sense of whether something is repairable and in what possible ways it might be fixed. In reality, such intuitions *do* rely on fairly complex cognitive processes, but mathematicians and plumbers are not consciously aware of them. Practiced natural intuitions are facilitated by training and expertise.

How can we explain the occurrence of intuitions in practiced natural capacities? As these skills are not the spontaneous outputs of intuitive ontologies but rely on highly culture-specific domains of knowledge, intuitions in such domains may seem puzzling at first. However, at least some cases of intuitions in practiced natural skills can be explained as the result of a redeployment of maturationally natural capacities. As we have seen, evolved cognitive capacities such as discriminating visual differences can be redeployed in a variety of cultural domains, such as chess, reading, and music. David Huron (2006, chapter 3) presented American and Balinese musicians with the opening notes of an unfamiliar Balinese gamelan piece. They were asked to guess which note would be next in the sequence. The weakest Balinese listeners were better than the best Americans, presumably an effect of cultural exposure, but interestingly the latter still performed significantly above chance. Apparently, they were able to quickly adapt to the unfamiliar tonal structure by relying on their expertise as Western musicians. Their steep learning curve was probably facilitated by cognitive predispositions that underlie musical scales. Since humans can discriminate between about 240 pitches over an octave in the mid-range of hearing (Fastl and Zwicker 2007), in principle a very large number of scales are possible. Yet, in practice, musical traditions only explore a modest subset of these possibilities: most scales across times and cultures contain only between five and seven tones, often with well-defined intervals between them. Gill and Purves (2009) argue that this conformity can be explained by the fact that such scales provide the best fit with human vocal fold vibrations, to which our ears are well attuned. The American musicians probably drew on these maturationally natural acoustic capacities to predict the direction the Balinese pieces would take.

Although the distinction between intuitive and reflective beliefs provides a useful and relevant characterization of beliefs, there may be some drawbacks. First, the term intuitive as it is commonly used in cognitive science has at least two senses. The first sense is related to the dual processing account: a belief is intuitive if it is (or could be) the output of maturationally natural cognitive processes such as intuitive ontologies. The second sense is more in line with the natural philosophical sense as it was understood by Descartes and is still used in philosophy today: intuitive beliefs are not the result of a reasoning process; we do not consider their source, nor their justification. Do these two senses have the same extension, that is, are beliefs that are spontaneous and unreflective always and only the outputs of maturationally natural processes? While some authors (e.g., Horst 2013) have attempted to provide a unified definition that combines these features, the existence of intuitions in expert domains makes this doubtful: they are not the outputs of maturationally natural processes but yet require no conscious deliberation or justification.

Second, the distinction between intuitive and reflective beliefs is dichotomous and therefore does not admit of degrees. Yet, some reflective notions seem to be better compatible with our intuitive beliefs than others. For example, although medieval impetus physics was a proto-scientific theory, it is more compatible with our intuitive physical notions than Newtonian or particle physics. Similarly, creationism is a body of reflective beliefs, but it is arguably more compatible with our untutored intuitions of teleology and design than evolutionary theory is (Blancke and De Smedt 2013). The distinction between intuitive and reflective beliefs is useful because it enables us to understand how we can hold mutually conflicting commonsense and scientific notions of the world, but as it stands, it is rather coarse-grained. This has led to a polarization in CSR about the relative intuitiveness of religious beliefs and the nonintuitiveness of science and theology. Indeed, as we saw in the Balinese music example, some skills are a combination of intuitive and reflective reasoning, and of practiced and maturationally natural skills. The same may be true of religious beliefs.

Many CSR authors have characterized religious beliefs as intuitive. For example, Kelemen characterizes children as *intuitive theists* (see chapter 4), and Bloom (2007) argues that humans have an intuitive mind–body dualism. By contrast, others have emphasized reflective aspects of religious beliefs. Ilkka Pyysiäinen (2003) maintains that religion depends critically on our ability to make representations of representations: religious beliefs are often bracketed off from other beliefs in a manner similar to other reflective attitudes, such as counterfactual scenarios or representations of other people's false beliefs. Even the most sincere belief in divine providence rarely leads people to leave everything up to God; believers will still act in a way that furthers their own goals. After all, the pious farmer does not stop tilling the soil, even though he believes the gods help his crops grow. Similarly, Boyer and Bergstrom write,

Most religious statements are represented not as simple propositions but as complex formulae of the form "Proposition *p* is *x*," in which the "*x*" may stand for "true," "guaranteed by the ancestors," "said by the prophets," etc. Such statements are me-ta-represented. ... They constitute "reflective beliefs," distinct from intuitive beliefs ... reflective beliefs are explicitly represented along with comments on their validity. (Boyer and Bergstrom 2008, 118)

Justin Barrett (2004, 10) also remarks that both intuitive and reflective beliefs play a role in the formation of religious concepts. Intuitive beliefs include "God has desires" and "God perceives human actions" (both from the domain of intuitive psychology), whereas reflective beliefs include "God exists as three persons" and "There is only one God."

In light of this, the claim that religious beliefs and practices are natural and intuitive requires some qualification. In many instances, they are a combination of both maturational and practiced natural capacities. Consider glossolalia, one of McCauley's exemplar cases for the naturalness of religion. Speaking in tongues relies strongly on maturationally natural cognitive capacities. Even so, the development of this skill requires some cultural exposure and practice. Glossolalia relies on our ability to produce and listen to natural language (a maturationally natural skill) but exploits the language faculty in a specific and unusual way (a practiced natural skill). In her spiritual autobiography, Lauren Winner, who was Jewish and converted to Episcopalianism in adulthood, recounts how she wanted very much to speak in tongues and prayed fervently to God for this gift of the Spirit. Nevertheless, she was unable to produce any glossolalia (Winner 2006, 257–258). Not having grown up in or having much familiarity with Pentecostal Christianity, Winner simply lacked the practice and cultural exposure to speak in tongues.

Religious beliefs and practices require cultural input for their development, more so than exemplar maturationally natural skills like walking and chewing. Children obviously do not reinvent or spontaneously come up with full-blown religious beliefs such as those found among practicing Christians or Hindus. But even at a more elementary level, there is little empirical support for the widespread idea that appeal to supernatural entities is the cognitive default. To the contrary, experimental evidence across cultures indicates that supernatural modes of explanation increase with age. Jay Wenger (2001) presented preschoolers, third-graders, and adults with difficult-to-explain scenarios and found that older children and adults were more likely to invoke God as an explanation than preschoolers. Cultural input is also important since participants from religious households mentioned God more than those from secular homes. Supernatural explanations thus do not always appear prior to naturalistic explanations, nor are they primitive or immature ways of thinking that get suppressed in the course of development. Rather, both modes of explanation (natural and supernatural) rely on cultural exposure, and in many cases coexist. Cristine Legare and colleagues found that children and adults across cultures combine natural and supernatural explanations to account for phenomena like the origin of species, illness, and death. For example, South Africans are aware of the biological causes of AIDS but nevertheless additionally accept witchcraft as a cause of AIDS in particular individuals: "A witch can put you in the way of viruses and germs" (Legare et al. 2012, 783). Such findings are not easily reconciled with the unqualified view that religion is natural. Rather, we can expect that religious beliefs and practices are an integration of maturationally natural skills (e.g., agency detection), practiced natural skills (e.g., speaking in tongues), and reflective attitudes (e.g., appeal to witchcraft as an explanation for disease).

This more qualified understanding of the terms natural and intuitive will be helpful for assessing the naturalness of natural theology in the remainder of the book. Since many religious beliefs and practices turn out to be combinations of maturational and practiced natural capacities and of intuitive and reflective attitudes, we will argue that the same is true of natural theology. In the next chapters, we will see that although natural theological concepts are sophisticated and the result of centuries of reflective reasoning, the intuitions that prompt and motivate them are often maturationally natural.

Summary

We have reviewed some theoretical assumptions that underlie CSR. Many cognitive scientists of religion conceptualize human cognition as composed of distinct domains of knowledge (intuitive ontologies) that arose through natural selection. These provide humans with an early developed set of inference mechanisms about physics, psychology, biology, and engineering. CSR scholars have applied this framework to understand cross-culturally recurring patterns in religious beliefs and practices.

A relative consensus in CSR holds that religious beliefs are natural and intuitive, but scholars differ in their interpretation of these terms. A closer examination of the cognitive psychological and philosophical meanings of naturalness and intuitiveness reveals that it is impossible to draw a categorical distinction between beliefs that are natural and those that are not, or between beliefs that are intuitive and those that are not, especially when we consider complex cultural practices, such as speaking in tongues. Nevertheless, we can look for the contribution of maturationally natural skills and intuitive beliefs within particular belief systems and cultural practices. The next chapter will look at the cognitive basis of our intuitions about divine attributes, keeping this nuanced understanding in mind.

3   Intuitions about God's Knowledge: Anthropomorphism or Preparedness?

The sixteenth-century reformer John Calvin ([1559] 1960, 35) noticed an intimate connection between our concept of supernatural agents, such as God, and normal agents, in particular, other humans: "Nearly all the wisdom we possess, that is to say, true and sound wisdom, consists of two parts: the knowledge of God and of ourselves. But, while joined by many bonds, which one precedes and brings forth the other is not easy to discern." The nature of this relationship remains obscure: does our thinking about other human beings inform our formation of religious concepts, or is it rather the other way around, that we start out overattributing superhuman powers and knowledge to others and downgrade it over time?

This chapter examines whether beliefs about divine attributes such as omniscience are maturationally natural. There are two theoretical approaches to the development of human knowledge of divine attributes: anthropomorphism and preparedness. According to the *anthropomorphism hypothesis*, divine attributes such as omniscience, omnipotence, and eternity are not intuitive and have to be acquired through cultural learning. By contrast, the *preparedness hypothesis* holds that young children's minds are intuitively attuned to such exceptional properties, and our initial stance toward others is one of overattributing knowledge and power. In this chapter, we will examine how humans intuitively conceptualize knowledge, exploring these competing explanatory frameworks. We will consider a recent dual-processing theory of mental state attribution to try to resolve this tension, arguing that anthropomorphism and preparedness each correspond to distinct maturationally natural modes of reasoning.

Is Natural Theology Cognitively Unnatural?

Religion and (natural) theology have divergent practices and scopes. Religions involve a rich array of habits such as praying, performing rituals, dressing appropriately, and observing food taboos, which are interwoven with many aspects of human life, such as sexuality, morality, and knowledge of the world. Theology, by contrast, is relatively sparse in its practice: it primarily involves reading, writing, and reflection. Whereas folk religious beliefs are acquired relatively easily, theology requires years of training, study, and deliberate practice.

CSR scholars have argued for a dichotomy between religion and theology: religion is natural; theology is not. McCauley, for instance, holds that theology is more akin to science than it is to religion:

Like scientists, theologians occupy themselves with forms of reflection that are difficult to learn and difficult to master and that occasionally even issue in representations that are just as cognitively unnatural. Theology is one of the few academic undertakings that can result in formulations that are very nearly as distant from and as obscure to humans' common understandings of the world as the most esoteric theoretical proposals of science are. (McCauley 2011, 212)

Theology likewise requires considerable formal instruction and study, quite a contrast to the effortlessness with which children acquire folk religious beliefs. Barrett (2011, 139) agrees that "while general religious thought and action have largely natural cognitive foundations, theology typically does not have anything like the same naturalness."

Two related concepts clarify this distinction between the cognitive underpinnings of religion and theology: theological correctness and theological incorrectness. With *theological correctness*, Barrett (2011, 138) denotes our reflective notions of theological concepts. Like political correctness, we have a reflective notion of which ideas are in line with theology, but when we are caught off guard, these are replaced by more intuitive but less theologically correct beliefs. *Theological incorrectness* occurs when religious believers unwittingly distort the official doctrines of their belief systems to fit their intuitive expectations. Jason Slone (2004) argues that theological incorrectness is the norm: believers who have some familiarity with theological doctrine will inevitably deviate from what theologians propose. For example, the culturally ubiquitous belief in luck is incompatible with theological systems where God knows and foresees everything (e.g., Calvinism), but this does not decrease folk beliefs in luck held by Calvinists. Likewise, Buddhism attributes real-life situations such as financial success or illness to karma, but in everyday life, Buddhists degrade karma to luck or in some instances regard karma as a personal (rather than an impersonal) agent that actively rewards or punishes them. Such instances of theological incorrectness constitute what Boyer (2002, 285) has termed "the tragedy of the theologian": in spite of the considerable effort theologians put in constructing coherent religious worldviews, these will always be distorted by the lay audience.

McCauley (2011, 212) cites the near absence of theology in non-Western, nonliterate cultures; theology is a practice for the select few working in highly institutionalized environments. However, a more detailed look at theological systems in non-Western cultures suggests that this view should be qualified. Even societies with little social stratification have some division of cognitive labor, with priests, shamans, or healers as religious specialists. Small-scale non-Western cultures have come up with sophisticated theological arguments and conceptions. To take but one example, consider the theology of the Kapauku, a horticultural, nonliterate society in the highlands of western Papua New Guinea. They have little material culture and a loose political structure. Yet, they have a well-developed theology describing a supreme creator being, Ugatame (the origin and direct cause of everything in the world, including good and bad), and his relationship to the world. As in philosophical theology in large-scale societies, the Kapauku not only systematically reason about the properties of their supreme being, they also make inferences about how he relates to the world. The anthropologist Leopold Pospisil recounts how an old man queried him about Christianity:

He could not understand how it was possible that the white man could be so clever and ingenious in designing such amazing contrivances as aeroplanes (which the old man could see flying over his valley), guns, medicines, clothes, and steel tools, and at the same time could be so primitive and illogical in his religion. "How can you think," he argued, "that a man can sin and can have a free will, and at the same time believe that your God is omnipotent, and that he created the world and determined all the happenings? If he determined all that happens, and (therefore) also the bad deeds, how can a man be held responsible? Why, if he is omnipotent, did the Creator have to change himself into a man and allow himself to be killed (crucified) when it would have been enough for him just to order men to behave?" ... Furthermore, the Christian notion of man resembling

God in appearance appeared to him as utterly primitive (*tabe-tabe*, "stupid").
(Pospisil 1978, 85)

This example indicates that literacy and research-oriented institutions,
while conducive to theology, are not necessary for its cultural development.

Divine Attributes

A recent challenge to the near-consensus that theology is unnatural is
the preparedness hypothesis, developed by Barrett and Richert (2003),
who argue that belief in divine attributes, such as omniscience, is maturationally natural. A large part of historical natural theology was concerned
with discovering what God was like. For example, although Paley's ([1802]
2006) *Natural Theology* is best known as a long, cumulative design argument for the existence of God (see chapter 4), it is also concerned with
inferring divine attributes from nature, as indicated by its full title: *Natural
Theology: or, Evidences of the Existence and Attributes of the Deity, Collected
from the Appearances of Nature*. The God of the Abrahamic religions is traditionally regarded as a personal being who is eternal, immutable (unchang-
ing), omnipotent, omniscient, free, and morally perfect. These properties
have been extensively discussed in Christian, Jewish, and Islamic natural
theology. For example, if God is morally perfect, how can we explain evil?
Can God's omnipotence and omniscience be reconciled with human free
will? Such topics clearly require reflective reasoning and philosophical
proficiency, but it is unclear how intuitions about divine attributes originate. Are they the result of practice, analogous to the practiced intuitions
of mathematicians or plumbers, or do they arise early in development,
analogous to maturationally natural intuitions about physical objects and
agents?

At first blush, an omnipotent, omniscient agent is unlike the other
agents we encounter in our ordinary experience. In chapter 2 we saw that
children come to understand persons as agents who act intentionally based
on mental states like desires and beliefs. Mature intuitive psychology takes
into account that agents' beliefs are limited, that is, that they can have
false beliefs. Our intuitive physics, likewise, conceptualizes agents as limited in space and time, and subject to physical constraints. Our intuitive
biology regards all organisms, including humans, as being born and subject
to growth and decay (more on this in chapter 8). Intuitive ontologies thus

seem to promote an understanding of agents as beings that are psychologically, physically, and biologically limited. Still, the belief in agents that
have special properties such as exceptional perceptual capacities, super-
powers, and privileged access to information is cross-culturally widespread.
Stories and myths often portray beings who are more powerful and knowledgeable than ordinary mortals, such as gods, culture heroes (e.g., Odys-
seus, Gilgamesh), and, more recently, superheroes. CSR scholars do not
agree on how beliefs in extraordinary agents develop and how these beliefs
relate to our intuitive understanding of normal agents. Proponents of the
unnaturalness of theology (e.g., Boyer 2002; Slone 2004) hold that divine
attributes are cognitively unnatural to represent. However, as we shall see
below, a minority of CSR scholars (e.g., Barrett and Richert 2003) regards
divine attributes as the cognitive default. This latter view is clearly more
germane to our hypothesis that natural theology rests on cognitively natural intuitions but seems difficult to reconcile with the idea that intuitive
ontologies promote an understanding of limited agents, which lies behind
the widespread idea of theological incorrectness.

God Concepts as a Form of Anthropomorphism

According to the anthropomorphism hypothesis, our concepts of God
and other supernatural beings are modeled on how we conceptualize ordinary human beings. This hypothesis goes back at least to Euhemerus, who
believed that the gods were deified renowned ancestors and rulers.[1] Piaget
(1929, part 3) proposed that young children acquire their religious beliefs
by modeling their concept of God on their parents and other adults. For
instance, they use the observation that other people create objects as a starting point to appraise God's properties as a creator (artificialism). Piaget's
position is interesting because he thought children initially tend to over-
extend power and knowledge to people, especially their parents, and that
later, through experience, they adjust their expectations. They attribute to
God the properties of omniscience and omnipotence that they first attributed to their parents. Stewart Guthrie (1993, chapter 7) outlines a detailed
hypothesis on the anthropomorphic origins of religious beliefs: the salience
and importance of human agents in our evolutionary history has led to a
tendency to overattribute human agency to natural events, leading to the
invention of supernatural agents; for example, rustling in the foliage and

gurgling water are interpreted as caused by anthropomorphic agents, leading people to posit the existence of sylvanian and riverine spirits.

Barrett and Keil (1996) were the first to empirically test the anthropomorphism hypothesis. They hypothesized that our inferences about humans
guide our thinking about God, especially when we have to infer what God
would do or know in concrete situations. They first asked Christian participants a series of questions that probed their understanding of divine properties, for example, whether or not God could attend to more than one event
at the same time. The subjects responded by giving the theologically correct
answer: God can indeed do so because he is omniscient, omnipotent, and
omnipresent. However, when they were asked to recall stories about God,
they inadvertently distorted them to fit God's behaviors and thoughts into
intuitive expectations about normal people; for example, they were likely
to misremember the stories to the effect that God could only have attended
to one event at a time. One of the stories went as follows:

A boy was swimming alone in a swift and rocky river. The boy got his left leg caught
between two large, gray rocks and couldn't get out. Branches of trees kept bumping
into him as they hurried past. He thought he was going to drown and so he began
to struggle and pray. Though God was answering another prayer in another part of
the world when the boy started praying, before long God responded by pushing one
of the rocks so the boy could get his leg out. The boy struggled to the river bank and
fell over exhausted. (Barrett and Keil 1996, 224)

When asked to reproduce this story, participants tended to misremember
it, stating that God first had to finish answering one prayer before he could
attend to the boy's. A later study found that Hindu believers similarly tend
to impose implicit limitations on Viṣṇu that are not endorsed in Hindu
theology (Barrett 1998).

If anthropomorphism is the cognitive default, we can expect that it is
present early in development and decreases to some extent as a result of
religious instruction. To test this hypothesis, Andrew Shtulman (2008) gave
adults and children a series of adjectives (e.g., skinny, bored, heavy, young)
that could be applied to humans but not to God. He asked participants
if these adjectives could be applied to religious beings (e.g., God, angels)
and imaginary beings whose existence is not generally endorsed (e.g.,
zombies, fairies). He found that adults do attribute human properties to
religious entities, but to a lesser extent than they attribute such properties
to imaginary beings. By contrast, five-year-olds attributed as many human

properties to religious beings as they attributed to fictional ones. Shtulman interprets the results as follows: over time, adult beliefs about supernatural beings become more theologically correct (as indicated by their explicit reference to theologically correct beliefs such as "God has no body," "God is everywhere"), whereas children are still anthropomorphizing.

Indirect support for the anthropomorphism hypothesis can also be found in scripture and depictions of gods. Many religious beings are represented in anthropomorphic form, for example, Hindu gods, Western African ancestors, and Celtic, Germanic, Roman, and Greek deities. Even cultures with a strictly non-anthropomorphic monotheism and an explicit prohibition on depicting God frequently use anthropomorphic imagery in their textual descriptions of God. The Hebrew Bible contains several anthropomorphic passages. For instance, after having eaten from the forbidden fruit, Adam and Eve "heard the sound of the Lord God walking in the garden at the time of the evening breeze" (Genesis 3:8); "I [God] will cover you [Moses] with my hand until I have passed by; then I will take away my hand, and you shall see my back; but my face shall not be seen" (Exodus 33:22–23); "I saw the Lord sitting on a throne, high and lofty; and the hem of his robe filled the temple" (Isaiah 6:1). In the Qur'an we find *surahs* such as "God said, 'Iblis, what prevents you from bowing down to the man I have made with My own hands?'" (38:75); "The East and the West belong to God: wherever you turn, there is His Face" (2:115).[2] Even more striking is the *ḥadīth* of the young man (*al-Asma' wa al-Sifat*), "The Messenger of God said: 'I saw my Lord under the form of a young man, beardless with curly hair, and clothed in a green costume.'"[3]

These anthropomorphic passages puzzled theologians for centuries. For instance, the medieval Jewish theologian Mosheh ben Maimon (Maimonides) devoted the first nine chapters of his *Guide for the Perplexed* to account for the frequent anthropomorphic imagery, insisting that it had to be understood figuratively:

Since men of greatness and authority, as, e.g., kings, use the throne as a seat, and "the throne" thus indicates the rank, dignity, and position of the person for whom it is made, the Sanctuary has been styled "the throne," inasmuch as it likewise indicates the superiority of Him who manifests Himself, and causes His light and glory to dwell therein. ... This is the idea which true believers should entertain; not, however, that the Omnipotent, Supreme God is supported by any material object; for God is incorporeal ... how, then, can He be said to occupy any space, or rest on a body? (Maimonides [12th c.] 1910, 21)

In Islamic philosophical theology, the interpretation of anthropomorphic imagery led to disputes between different religious schools, including
the anthropomorphists (*mushabbiba*), who thought this imagery should be
taken literally, the Mu'tazilites, who believed it should be interpreted meta-
phorically, and the Ash'arists, who regarded it as neither (El-Bizri 2008). In
Christian theology, there continues to be discussion on whether a lack of
anthropomorphic features would place limits on God's power and knowledge, and hence a denial of omnipotence. After all, an incorporeal being
cannot ride a bicycle or eat ice cream (Gleeson 2010). But Linda Zagzebski
(2008) claims that God can still know what it is like to experience these
things without having to do them.

The anthropomorphism hypothesis has implications for the naturalness
or unnaturalness of natural theology. If it is correct, it seems that at least
that part of natural theology that is concerned with divine attributes is
unnatural, since the cognitive default would be to think about other agents
as limited in their capacities and knowledge. Formal religious education
can encourage adults to attribute "radically nonhuman, counterintuitive
capacities to deities (e.g., total omniscience)," but "in everyday judgment
adults tend to think of deities in terms that are more human-like and limited" (Lane, Wellman, and Evans 2010, 1476).

The Preparedness Hypothesis

Barrett and Richert (2003) have proposed the preparedness hypothesis as an
alternative to the anthropomorphism view. *Preparedness* is a standard term
in evolutionary biology and psychology: a cognitive system is prepared
for x if it is—through evolution—predisposed to represent x or to respond
to x in a certain way. For example, evolutionary psychologists hold that
humans are prepared to be fearful of snakes, spiders, and darkness, but not,
for instance, of cars and guns, because the latter, unlike the former, played
no role in our evolutionary history.

According to the preparedness hypothesis about divine attributes,
reasoning about God initially structures reasoning about human beings:
"Early-developing conceptual structures in children used to reason about
God are *not* specifically for representing humans, and, in fact, actually
facilitate the acquisition and use of many features of God concepts of the
Abrahamic monotheisms" (Barrett and Richert 2003, 300). This is because

children's cognitive equipment assumes that "many superhuman properties are the norm" (Barrett and Richert 2003, 303). God concepts, especially
of the Abrahamic religions, play into these default assumptions.

This hypothesis is surprising in several ways: it not only subverts the
dominant assumption that God concepts are anthropomorphic but also
the widespread view (e.g., Hume 1757) that polytheism is more cognitively
natural than monotheism. Indeed, monotheism requires specific cultural
conditions such as a large-scale society with economic differentiation, literacy, a sedentary mode of subsistence, and political inequality (Sanderson
and Roberts 2008).

The preparedness hypothesis relies on experimental observations that
indicate that children tend to overattribute power and knowledge to others, especially to adults. The best-studied divine attribute for which CSR
proposes intuitive knowledge is omniscience. Omniscience is usually
understood as perfect knowledge. A person who is omniscient has only true
beliefs and no false beliefs. Or, differently phrased, for every proposition p,
an omniscient person knows that p is true or that p is false (Wierenga 2009).
Some authors (e.g., van Inwagen 2006) have made the stronger claim that
an omniscient person not only has no false beliefs, but also that it would be
metaphysically impossible for that person to believe anything that is false.
According to the preparedness hypothesis, it is more cognitively natural
to represent agents that have only true beliefs than agents that have both
true and false beliefs. As a consequence, children younger than the age of
four or five tend to think about other agents as having only true beliefs.
Strictly speaking, this concept is not the same as the theological concept of
omniscience since an omniscient agent not only knows some true beliefs
but all true beliefs. Yet even in this restricted, psychological sense, it seems
remarkable that children would have this intuition.

On the face of it, omniscience seems like a counterintuitive property.
In our ordinary experience, agents are limited in their knowledge, and
they can hold false beliefs. However, in some respects, omniscience might
be easier to represent than being cognitively limited. For one thing, one
need not keep track of an omniscient person's false beliefs since her beliefs
always coincide with reality. In order to attribute a false belief to a subject S,
we need to make a distinction between the state of the world and S's mental
representation of it. This allows us to make sense of and rationally explain
a variety of S's actions. For instance, suppose S makes a long detour to get

to work even though the roadblock that previously prompted her to make this detour was lifted yesterday. This puzzling behavior can be adequately explained by supposing that S is not aware that the roadblock is no longer in place. S has a false belief, which explains her actions. Suppose now that S is omniscient: in this case, the state of the world and her mental representations overlap, and S will always act in accordance with the state of the world. Hence, it would be easier to process beliefs of an omniscient being than of a being who is limited. Indeed, we can simply think about the reality of a situation without attributing any mental states at all, a tendency in young children termed *reality bias*. Accordingly, young children start out by assuming that everyone's knowledge corresponds to the state of the world and only later make fine-grained distinctions. Their initial beliefs about God's knowledge are thus more correct than their beliefs about what other agents know.

Empirical support for Barrett and Richert's claim that attributing omniscience is maturationally natural comes from a modified false belief task by Barrett, Richert, and Driesenga (2001). As we saw in chapter 2, cross-cultural studies indicate that children younger than the age of four or five have difficulties attributing false beliefs to other agents. For example, when they are shown that a cracker box contains rocks, they will incorrectly infer that other people, including those who have never seen the contents of the box, will believe that there are rocks in the box. Barrett et al. (2001) hypothesized that children start out by attributing super-knowing properties to everyone and later adjust their expectations. If this is the case, we should see a change with increasing age in the beliefs children attribute to others: most three-year-olds will incorrectly predict their mother believes that there are rocks in the box, whereas five-year-olds will attribute to her the false belief that there are crackers inside. However, given that children receive testimony that God is omniscient, there will be no change in their attribution of true beliefs to God: young children as well as older ones will think God believes that there are rocks in the box.

To test their hypothesis, Barrett et al. (2001) presented three- to six-year-old Protestant children with a typical cracker box. They asked them what was inside; most children guessed crackers. The experimenter then revealed that the box contained small rocks. Participants had to assess what a variety of agents (their mother, God, a bear, an ant, and a tree) would believe was in the box. In line with their expectations, the researchers found that most of the youngest children predicted that all agents would believe that there were rocks inside. Older children became increasingly aware that the other agents—but not God—would think that there were crackers in the box. The experimenters interpreted these findings as incompatible with anthropomorphism: if children modeled their concept of God on other agents, they would start out thinking that everyone is fallible and only later learn through testimony that God is not subject to these cognitive limitations. But the reverse held: they started out with attributing omniscience to everyone and at a later age restricted this to God. Similar results were obtained with Roman Catholic Maya children (Knight et al. 2004).

Barrett and Richert (2003) argue that children's minds are naturally attuned to believe in the God of the Abrahamic monotheistic religions, a puzzling interpretation. We would expect that our minds are shaped by natural selection to reason about agents like us and not about divine agents. Why would natural selection endow humans with an incorrect notion about the properties of other agents, which needs to be adjusted through experience? Moreover, preparedness seems difficult to reconcile with the theories and evidence from CSR that indicate that we frequently anthropomorphize gods.

A Conflict between Anthropomorphism and Preparedness?

On the face of it, it seems that anthropomorphism and preparedness are incompatible. But remarkably, one author (Justin Barrett) has defended both positions, anthropomorphism (Barrett and Keil 1996, Barrett 1998) and preparedness (Barrett and Richert 2003). The latter study briefly discusses this apparent tension but explains the discrepancy as follows. Young children start out with a theologically correct concept of agency, but if in adulthood this concept is not available or not salient, adults will use more anthropomorphic concepts to fall back on: "Indeed, especially when adults are generating non-reflective, real-time inferences about God they may unwittingly use a concept of God that includes anthropomorphic properties they reflectively reject" (Barrett and Richert 2003, 310). This reply is not satisfactory because it fails to explain why anthropomorphism occurs in scripture as well and why a number of monotheistic (Muslim and Christian) theologians have accepted some anthropomorphic properties of God. Also, if the divine attributes are so intuitive that even three-year-olds can reason about them, it seems implausible that fully enculturated adults would not be able to rely on them when they are not salient.

Lane et al. (2010) have attempted to resolve the problem by explaining preparedness about omniscience as an experimental artifact. Their study makes finer-grained distinctions between age groups, finding that, during a brief period (around four years) children do attribute false beliefs to God. Similarly, they are more likely to attribute ignorance to God during this period. Lane et al. interpret these findings as follows: children start out with a reality bias when they assume that God's mental capacities are similar to those of others. Around age four, they learn to attribute false beliefs, which they extend to God, and only later, at around five years of age, they discriminate between God and other minds. These findings are in accordance with anthropomorphism, since God is conceptualized similar to other agents early in development and the distinction between God and other agents only arises later, presumably through formal and informal religious instruction. At first sight, this seems to result from the same belief as at age three, but the cognitive mechanism underlying this belief is different: at age three, it is the reality bias, whereas five-year-olds are enculturated in the belief that God is all-knowing while fully realizing that other agents do have false beliefs. Lane et al.'s (2010) interpretation does not explain away preparedness, since they do not account for the initial reality bias. Even if children attribute false beliefs to God during a short phase of their developmental trajectory, it still seems surprising that the younger ones attribute omniscience to everyone.

Toward an Integrated Account

Anthropomorphism and preparedness provide us with two conflicting hypotheses, each supported by empirical studies. Under anthropomorphism, representing beings with unlimited knowledge is nonintuitive, whereas under preparedness it is intuitive. Both hypotheses implicitly assume a rich and composite notion of cognitive naturalness. Accordingly, maturationally natural skills are regarded as fluent, fast, requiring relatively little conscious effort, and supporting many inferences. McCauley (2011, chapter 1) provides a beautiful illustration of this composite notion. He relates an encounter between two characters in *Barchester Towers*, a nineteenth-century novel by Anthony Trollope. Mr. Slope, a chaplain working for the local bishop, has earlier promised to provide Mr. Quiverful, vicar and struggling father of fourteen children, with a substantial and handsome new position. However, due to political machinations, Mr. Slope has to retract this offer and comes to tell Mr. Quiverful the bad news. It is striking how Trollope describes that Mr. Quiverful notices at a glance that his hopes for the position are dashed the moment Mr. Slope begins to speak to him:

As he spoke, the poor expectant husband and father saw at a glance that his brilliant hopes were to be dashed to the ground, and that his visitor was now there for the purpose of unsaying what on his former visit he had said. There was something in the tone of the voice, something in the glance of the eye, which told the tale. Mr. Quiverful knew it all at once. (Trollope [1857] 1994, 188)

The incident is indeed striking, but the question is whether this cognitive feat is exemplary for all instances of maturationally cognitive processes, as McCauley would have it.

In many cases, there is a trade-off between inference-richness and flexibility on the one hand, and speed and efficiency on the other. Cognitive processes that are quick and efficient are typically less flexible and inference-rich. Take, for example, our capacity to detect eye-gaze direction. This capacity emerges developmentally early and is very quick and efficient but relatively inflexible. It misfires frequently, as when we detect staring eyes not only in other agents but also in stains on walls and carpets (Guthrie 1993). By contrast, folk taxonomy provides rich inferences. It is the output of folk biological reasoning that emerges robustly in childhood in a variety of cultural conditions. It is flexible, as it is sensitive to distinct ecological conditions: children who grow up in rural South America typically develop a richer folk taxonomy compared to children from the urban areas of the United States. Children from urban environments often have impoverished taxonomies; for example, they regard deciduous trees as "trees" rather than as basic-level taxonomic entities like oaks and beeches. This is because folk taxonomy requires cultural support and learning (Atran, Medin, and Ross 2004). Because flexibility requires input of specific environmental and cultural conditions, there is a trade-off between speed and flexibility. Folk taxonomy is slow but flexible, whereas eye-direction detection is fast but inflexible. Given functional incompatibility (see chapter 2), we can expect that such diverging demands call for different cognitive solutions. Is this the case for mental state attribution?

Ian Apperly (2011) proposes that there are at least two systems involved in mental state attribution: *low-level mind-reading*, which is fast, implicit, and inflexible, and *high-level mind-reading*, which is slow, explicit, and flexible. We attribute mental states to others under a wide variety of conditions. In some cases, humans need to make quick and dirty inferences about what others perceive and know. For instance, when driving on a busy street, one needs to assess whether or not a cyclist intends to cross the lane. This requires paying attention to the environment and to the behavior of the cyclist. Based on a quick assessment of behavioral cues and traffic flow, one has to make a decision about the cyclist's intentions. As this system pays primarily attention to the agent and not to oneself, it is allocentric. This type of inference about other minds is fast, automatic, but relatively inflexible: we do not (and cannot) take into account the rich mental life and motivations of the cyclist.

In other circumstances, we do need to make complex, subtle inferences, for instance, when we have to engage in strategic negotiations or understand our own position in relation to others. Realizing that a friend suddenly behaves coldly and distantly calls for a search for causes that may have prompted this shift in behavior. Is she like this to everyone or just to me? Did I do anything to upset her? This type of inference is typically egocentric: we adopt the person's perspective to consider why she behaves the way she does. We gauge a variety of cues that allow for rich, flexible inferences, but the process is somewhat slow and effortful; for example, only after a while, it dawns on me that I have forgotten her birthday.

Both types of inference are maturationally natural according to Apperly. Adults can rely on either system, depending on the situation. The fast, inflexible system matures early and is already available to infants. The slower, more reflective system matures only between four to five years of age. We propose that the tension between preparedness and anthropomorphism in God concepts can be resolved if we assume that thinking about God's mental states also relies on these two systems of intuitive psychology. Preparedness is elicited by the slow, effortful system, whereas anthropomorphism is prompted by the fast, inflexible one.

Preparedness and Cognitive Load

According to the preparedness hypothesis, young children exhibit a reality bias when they reflect on what others believe. Interestingly, this bias is not restricted to children but also occurs in adults during experimental manipulations that increase cognitive demands. For example, Newton and de Villiers (2007) found that adults fail simple false belief tasks if they have to verbally "shadow" (i.e., repeat with a slight delay) a spoken text.

Even when we give a task our full attention, it is often difficult to keep in mind the mental limitations of others. Keysar et al. (2000) asked participants to move an object in response to the instructions given by a "director." Objects were placed in a grid-like cupboard, which had opaque as well as transparent wall segments. The director stood behind the cupboard and could thus only see the objects placed in the cells that had a transparent wall segment. Participants stood in front of the cupboard and could see everything inside it. They had to take into account what the director could see. For instance, if the director asked to move "the smallest candle," this was actually the second-smallest candle, since the smallest candle was placed in front of an opaque segment and thus only visible to the participant. Although subjects corrected themselves frequently and tried to keep in mind the visual perspective of the director, they made mistakes in about 20 percent of cases (e.g., moving the smallest candle)—a result in line with the reality bias exhibited by children younger than four years of age. In another study (Apperly et al. 2008), adults read pairs of short sentences. The first always described someone's beliefs (e.g., "He thinks the object on the table is red"), while the second always described reality. In some cases, the unrelated condition, the description of reality, differed from the belief of the agent but was not in conflict with it (e.g., "Really, the object on the chair is blue"). However, crucially, in other cases, reality was in conflict with the belief of the agent, the false belief condition (e.g., "Really, the object on the table is blue"). Subjects were faster and made fewer mistakes recalling the sentences in the unrelated condition than in the false belief condition.

In our everyday lives, we also sometimes forget that other people do not have access to the same information we do (e.g., "Oh, yes, I forgot, you didn't hear about that yet"). Why do adults sometimes fail to take into account mental limitations of others? Samson and Apperly (2010) propose that there are two factors that make explicit theory of mind tasks cognitively demanding: we need to resist our own knowledge about features of the environment, and we need to take into account what information is relevant for particular judgments. While false belief reasoning is maturationally natural, it requires cognitive effort.

Explicit false belief reasoning and knowledge attribution rely on a fluent, inference-rich, but slow system. Interestingly, the properties of this system suggest that it is easier to attribute knowledge or ignorance than true or false beliefs. A meta-analysis (Wellman and Liu 2004) indicates that children can indeed attribute ignorance to agents before they can attribute false beliefs to them: three-year-olds can successfully predict that their mother will not know what is inside a closed box that she has never seen before, but they cannot predict that she will have a false belief about its contents (as in the crackers versus rocks example). Although chimpanzees fail false belief tasks, several studies indicate that they are able to attribute ignorance to others. For example, chimpanzees preferentially select a reward that was hidden in their competitor's absence over one that was hidden in its presence (Kaminski, Call, and Tomasello 2008). One reason for these results might be that the "pull of the real" (i.e., reality bias) poses less of a problem for ignorance tasks than for false belief tasks. In false belief tasks where young children have to keep the agent's beliefs in mind, they sometimes become confused by their own knowledge. By contrast, ignorance tasks do not require one to keep mental states of the agent in mind that might conflict with one's own mental states: children do not have knowledge that can be confused with the agents'. Indeed, they can simply use their own lack of knowledge to extrapolate the knowledge that others may not have.

Children starting at about four years of age are able to make inferences about the knowledge of others, but as we have seen, such inferences require an explicit distinction between reality and belief. Equating knowledge with reality (reality bias) is reliably demonstrated in children younger than four and in adults who are put under high cognitive load. We can explain the preparedness findings as a result of this bias. Agents whose knowledge corresponds to reality (e.g., the Abrahamic God) are easier to reflectively think about than agents with false beliefs.

Anthropomorphism and Efficient Cognitive Processing

If our analysis is correct and the human attribution of omniscience indeed corresponds to a natural cognitive tendency (reality bias), how can the anthropomorphism findings be explained? Several recent studies indicate that adults and infants do not systematically exhibit reality bias under all circumstances but that they are sensitive to the mental states of others. This is especially the case in situations where they can easily infer the mental states of others from visual and other environmental cues. For example, Kovács et al. (2010) let adults perform a visual detection task. They saw an agent placing a ball on a table in front of an occluder. Then the ball rolled behind the occluder. The experimental situation was then manipulated so that the agent had either a true or false belief about the location of the ball (either the agent was there when it happened or he was not). Interestingly, adults were faster in detecting where the ball was when its location corresponded with the true belief of the agent, even though the agent's beliefs were irrelevant to this task. A similar procedure with infants of seven months old revealed comparable results. The infants looked longer at the ball when the agent's belief about its location turned out to be false, that is, when the agent had left prior to the ball rolling away. Apparently, infants compute other agents' beliefs and do not reason purely egocentrically.

A similar result was obtained by Qureshi, Apperly, and Samson (2010), who showed that adults immediately and involuntarily take into account the mental state of others, even when they have to focus on purely egocentric tasks. Subjects were shown pictures of a person standing in a room decorated with a small number of dots. In some cases the number of dots the participant could see was the same as the number of dots the agent could see, for example, the agent was facing a wall with two dots. But sometimes there was a mismatch, for example, the agent faced a wall with one dot, but his back was toward another dot; the participant, looking into the room, could see both dots. Participants were instructed to ignore the agent and just say how many dots they saw. Nevertheless, when the number of dots they saw was the same as that observed by the agent, they were significantly faster. They made more mistakes if the agent saw a different number, despite explicit instructions to just concentrate on the number of dots.

This indicates that next to the reality bias, we do take into account the perspective of others, especially when tasks are fast and require little reflection. Attributing beliefs to others, including false ones, is something we can automatically do, provided that there are rich contextual cues from which the agent's beliefs can easily be inferred. A disadvantage of this fast system is that it is inflexible: we cannot take into account prior knowledge of the agent, as we do in the slower and more explicit way of inferring beliefs. The anthropomorphism in Barrett and Keil (1996) and Barrett (1998) and the use of anthropomorphic imagery in scripture can be explained as the outputs of this fast, inflexible system. In these situations, God is a social agent in a rich context (e.g., Genesis 3:8–11, the stories in Barrett's experiments), and such rich contexts are likely to elicit inferences about limited agents with true as well as false beliefs. (It is interesting to note that God even asks where Adam is and who told him he was naked. It is clear that omniscience is not presupposed in this narrative.) Participants who heard the story about the swimming boy and learned that God was answering another prayer made fast mental state attributions about what God was attending to. Since this system is inflexible, they did not take into account God's omniscience, something they might have done when performing explicit, slow tasks.

This leads to the prediction that anthropomorphism occurs in narrative parts of scripture where God is interacting with others, as these interactions call into play our ability to attribute mental states, as in Genesis 3, where God is not explicitly present when Adam and Eve eat of the forbidden fruit. Given that he is not explicitly present, it is easy and natural to assume that he does not know what happened. By contrast, when authors of the books of the Bible are reflecting on God's beliefs, the slower, more flexible system comes into action, as in Psalm 139 (1–4):

O Lord, you have searched me and known me.
You know when I sit down and when I rise up;
you discern my thoughts from far away.
You search out my path and my lying down,
and are acquainted with all my ways.
Even before a word is on my tongue, O Lord, you know it
    completely.

This is where the attribution of omniscience comes into play.

The Tragedy of the Theologian Revisited

We can now look at the dichotomy that CSR scholars habitually draw between theologically correct/cognitively unnatural and theologically incorrect/cognitively natural. This dichotomy oversimplifies the complex cognitive processes that underlie religious thinking. Reasoning about divine omniscience is not cognitively unnatural or natural, simpliciter. Rather, these terms need to be qualified with respect to particular modes of representation and reasoning: omniscience is natural/intuitive as evaluated by a reality-biased system for belief-attribution, but unnatural/counterintuitive for one that attributes minds with limited cognitive access to the world. Scriptures and the writings of theologians reflect both modes of reasoning about God's mental states.

In theology, anthropomorphism has not altogether been shunned. As we have seen in Islamic theology, several schools of thought were congenial to the idea that God has hands, eyes, and other anthropomorphic attributes. Recent Western philosophy of religion also leaves room for anthropomorphism. For instance, Swinburne (2010, 10) has argued in favor of the view that God's powers are similar to our powers but amplified: "He moves the stars, as we move our arms, just like that—as a basic action." Given Swinburne's sophisticated writings on divine properties, it seems unlikely that this is a lapsus, a theologically incorrect idea slipping into an otherwise theologically correct framework. Similarly, William Hasker (2007, 156) states that "God is an agent, one who can and does act in particular times and places and with respect to particular configurations of persons, circumstances, and events." Not all theologians and philosophers of religion endorse this anthropomorphism. Maimonides promoted a negative theology where he argued for God's radical difference from his creation. Still, anthropomorphism recurs throughout theological history, and according to Andrew Gleeson, it has become very common in analytic philosophical theology over the past sixty years. He describes anthropomorphism as the view that "God, like human beings, is an agent acting in a 'punctiliar' manner, i.e., attending now to this, now to that, as circumstances arise (e.g., from the free choices of created beings) with the passage of time" (Gleeson 2010, 607). This describes very well how participants saw God's actions and knowledge in the studies by Barrett and Keil (1996) and Barrett (1998). Since this concept of God has also pervaded theology, it is problematic to call anthropomorphism "theologically incorrect." At best, we can call it a minority position within theology.

Why does anthropomorphism remain a viable option in philosophical theology? When thinking about God's knowledge and actions, especially his interactions with his creation, it becomes difficult to say anything meaningful without any anthropomorphic analogies. Rather than conceptualizing it as slippage into theological incorrectness, an explicit endorsement of anthropomorphism in philosophy of religion and theology provides a way to speak about God in an intelligible manner. These anthropomorphic terms can be used to describe what would otherwise remain epistemically opaque, inaccessible to our knowledge. In this way, speaking about God's hands, emotions, or interactions in time, is a way to make theological discourse more amenable to human cognitive predispositions, and this is probably why Swinburne and others write in this way about God.

Summary

Calvin observed a close connection between reasoning about God and thinking about humans. CSR scholars disagree on which is prior: does reasoning about divine minds and powers shape beliefs about ordinary humans (preparedness), or does reasoning about ordinary agents inform belief in supernatural agents (anthropomorphism)? This chapter has looked more closely at preparedness and anthropomorphism to unravel human reasoning about divine attributes, with a focus on omniscience.

We propose a third model, where reasoning about God's beliefs is subserved by two distinct systems in intuitive psychology: a slow, flexible system that favors representations of omniscient agents, and a fast, inflexible system that prompts humans to think of agents as limited and fallible. Both forms of reasoning are not only found in scripture, but also in theology: anthropomorphism constitutes an important minority tradition in medieval Muslim and contemporary Christian theology and philosophy of religion. This complexity in beliefs about God's attributes indicates that it is not always straightforward to denote what is theologically correct or theologically incorrect. Rather, maturationally natural inference systems continue to inform theology. This suggests more continuity between religion and theology than has been maintained in CSR.

4 Teleology, the Design Stance, and the Argument from Design

For many scientists and philosophers, Hume's philosophical critique (1779) and Darwin's theory of natural selection (1859) have undermined the idea that we can draw any analogy from design in artifacts to design in nature. Yet the argument from design stands as one of the most intuitively compelling arguments for the existence of a divine creator. It enjoys an enduring appeal, going back as early as Socrates (recorded in Xenophon's *Memorabilia*), Plato's *Timaeus*, and Cicero's *De Natura Deorum*. The design argument can also be found outside of Western religious traditions, for instance in Hinduism, with authors such as Śaṅkara and Udayana defending it. From the seventeenth to the nineteenth century, it was the most popular natural theological argument for the existence of God, with Paley as the best-known, albeit a late, example. Since the later decades of the twentieth century, the design argument has been experiencing a renaissance, especially in the form of the fine-tuning argument, which holds that the physical constants, the laws of nature, and the initial conditions of the universe are finely tuned to be life-permitting.

In what follows, we will examine empirical studies from developmental and experimental psychology to investigate the cognitive basis of the design argument. We will focus on two aspects: the tendency of humans to discern teleology in nature and the way they intuitively assess probabilities. A better understanding of these features not only elucidates the lasting popularity of the design argument but can also help theists and nontheists to construct a rational basis for disagreement.

The Argument from Design

The argument for the existence of a divine creator, based on evidence of design in nature, has recently enjoyed a revival in theology and philosophy (as exemplified by papers collected in Manson 2003). Even a hardened atheist like Dawkins (1986, 4–5) praises the argument for its explanatory coherence and intuitive appeal. Paley's image of the watch on the heath was certainly not the earliest formulation of the design argument, nor even of the watch analogy, but its familiarity makes it a suitable starting point. His argument can be summarized as follows: if one encounters a watch, the complexity of this artifact and the interrelations of its parts lead to the inference that it is the product of purposive design. From this, Paley ([1802] 2006, 7–15) concluded that complexity in nature is also the product of a designer because proposing that it could have been brought about by chance would be absurd.

Elliott Sober (2004, 118) formalizes Paley's design argument (and other arguments like it) as follows: there is an observation O and two possible hypotheses (H_1 and H_2) to explain it:

O: The watch has features, in particular, design, complexity, and a certain delicacy, that is, if its parts were a bit different, it would function less well or not at all. There are two hypotheses to explain these features:
H_1: The watch was created by an intelligent designer.
H_2: The watch was produced by a mindless chance process.

Paley relies on the intuition that O would be unsurprising if H_1 were true but highly surprising if H_2 were true. O therefore favors H_1 over H_2. Paley surmises that organisms are similar to watches in that they exhibit complexity, design, and fine-tuning. Therefore, running the same argument, H_1 is a better explanation for these features than H_2. From this we can conclude that organisms are created by an intelligent designer.

This way of arguing is not unusual. The Hindu philosopher Śaṅkara (8th c. CE) posited an intelligent, purposive craftsman (Brahman) as the originator of the universe. His design argument was intended to show that materialism, the view that the world originates from a non-intelligent, material first principle is a worse explanation for the apparent design of the world than creationism:

Consider that in ordinary life no non-intelligent entity is observed to produce modifications suitable for satisfying the purposes of some particular person, by itself, without being superintended by an intelligent agent. In ordinary life what we do see is that houses, palaces, couches, seats, pleasure-gardens, and the like, which are useful for obtaining pleasure and avoiding pain at appropriate times, are constructed (*racitā*) by intelligent craftsmen. In like manner, observe that this entire universe, externally consisting of the earth and other elements, is suitable for experiencing the fruits of various acts. ... Since even the most competent craftsmen cannot comprehend (the world's construction), how could the non-intelligent Material Nature (*pradhāna*) devise (*racayet*) it? In the case of such things as a lump of earth or a stone, no (power of contrivance) is seen, but the design (*racanā*) of special forms out of such things as clay is seen when they are superintended by potters and the like. In the same way, Material Nature (transforms itself) only when connected with a superintending, external intelligence. ... Therefore, since the design of the world is otherwise inexplicable (*racanā–anupapatti*), its cause is not to be inferred as non-intelligent. (Śaṅkara, cited in Brown 2008, 108)

These arguments from design have interesting epistemic properties: they are both an analogy and an argument to the best explanation. Analogies map the structure of a well-known domain (the source domain) onto a lesser-known problem (the target domain). In distant analogies, the structures of source and target domains greatly differ in their basic ontological properties. The design argument is a distant analogy in that it maps the artificial domain (source domain) onto the natural world (target domain). Artifacts exhibit goal-directedness in their design: they are intentionally created by designers who have their functions in mind. As organisms exhibit goal-directedness, they must also be products of a purposeful designer.

Although this analogical structure has often been attacked on the ground that it is inconclusive (e.g., Hume 1779; Frank 2004), it is worth noting that distant analogies are well-established in scientific practice as a way to gain insight into new problems (De Cruz and De Smedt 2010b). When the conceptual structure of the target domain is relatively unknown, as in the case of scientific discovery, scientists often resort to analogical reasoning. Historical examples include Johannes Kepler's mapping of the properties of gravity onto the properties of light (i.e., the fact that sunlight dissipates with increasing distance between the Sun and the planets it is cast upon) to explain why planets further from the Sun move more slowly, in this case, caused by a weakening of the gravitational force with increasing distance from the Sun (Gentner et al. 1997). A contemporary example is the Swiss army knife analogy as a way to conceptualize the evolved structure of the human mind in evolutionary psychology (Cosmides and Tooby 1994a, 88). The analogical structure of the design argument is thus epistemic; it is performed to gain insight into an unfamiliar domain, God's creation. In many early versions, the design argument served a heuristic, rather than a strictly argumentative purpose, as in the works of seventeenth-century natural philosophers such as Nieuwentijt, van Leeuwenhoek, and Ray, who perceived design as a source of wonderment about nature.

Traditionally, design arguments had an inductive argumentative structure. They began with the empirical observation that all complex, functional objects of known origin were products of intelligent design. Then came the inductive step, in which one infers that what is true for some members of a class is true for all members. Hume (1779, 56–60) dispensed with this way of reasoning by arguing that artifacts and biological organisms are too dissimilar to be classed together. When we see a house, we can reasonably infer that it has an architect or builder because we know from experience that this particular effect flows from that particular cause. But we have no guarantee that the universe bears such a resemblance to a house as to invoke a cosmic designer; we do not know in how far our analogy is reliable.

Although Paley did not mention Hume explicitly, his watchmaker analogy sought to escape this criticism by adopting a different strategy. It is an argument to the best explanation that relies on an inference to the best explanation (IBE), which has the following structure:

IBE Given evidence E and a pool of plausible, potential explanations $H_1, \ldots, H_n$ of E, if H_i explains E better than any of the other hypotheses, infer that H_i is closer to the truth than any of these others. (Douven 2002, 359)

IBE enables us to probabilistically infer that a given hypothesis is closer to the truth than other hypotheses because it explains the available evidence better than rival explanations. In this probabilistic aspect, the design argument differs from deductive proofs for the existence of God, such as Anselm's ontological proof (see Sober 2002; Swinburne 2004, for probabilistic discussions of the design argument). Arguments to the best explanation are meant to avoid Hume's critique by not relying on induction. Rather than having to make a positive case for why artificial design is analogous to divine design, they argue that there is no better explanation for order and complexity than design, regardless of where these properties are instantiated (Gliboff 2000). Having established apparent design and probabilistic inference as two key properties of the design argument, we will now examine their cognitive basis.

How We Infer Design

The Design Stance

To Paley, the conclusion that a watch is purposefully designed was self-evident. Our perception of its interrelated parts, formed and adjusted to each other—the coiled elastic spring, the flexible chains, the cogwheels—each fashioned out of the materials that suit their intended function best, should lead us to infer that the watch must have had a maker who formed it for a specific purpose. Yet Paley acknowledged that such seemingly spontaneous inferences require contextual knowledge about the artifact under consideration: "It requires indeed an examination of the instrument, and perhaps some previous knowledge of the subject, to perceive and understand it" (Paley [1802] 2006, 8). In the case of the watch, Paley could infer the intent of the designer, as he was familiar with the class of artifacts to which the watch belongs.

What would happen if Paley had pitched his foot against an unfamiliar object, such as a laptop? Would he have inferred design as automatically as in the case of the watch? Its sleek shape, carefully integrated buttons, and intended functions would have presented a puzzle to him. Unfamiliar artifacts can be so outlandish that people can be led to believe that they are not the work of human designers. From the nineteenth-century wave of European colonization onward, and especially during World War II, the indigenous inhabitants of Papua New Guinea were confronted with a cornucopia of Western goods. They believed that these were gifts of the ancestors. This led to the emergence of cargo cults, whose adherents are still trying to ritually lure airplanes into dropping more cargo, more Western goods. And what to think of people who observe UFOs? Often these alleged spaceships are no more than military reconnaissance airplanes, crashing weather balloons, or even bright planets. Nevertheless, people do infer that the objects are intentionally designed by supernatural agents and by extraterrestrials respectively. To gain a better understanding of the design argument, it is therefore useful to examine how humans infer design and how creator and artifact are causally linked.

Bloom (1996) has argued that manufacturing and understanding artifacts is governed by an intuitive design stance, a term he borrowed from Daniel Dennett (1987, 16–17). As we saw in chapter 2, humans take the design stance when they consider an object as designed to fulfill some function or as belonging to some class of objects. Remarkably, neither complexity nor order is a necessary or sufficient condition to decide whether an object is purposefully created. Gelman and Ebeling (1998) showed two-year-olds a stain vaguely resembling a bear. They told some subjects that the spot was created accidentally, by kicking over a bottle of paint, whereas another group learned that the shape was painted intentionally. Only the children in the latter group called it a bear. Thus the perception of the stain as accidental or representational depends critically on the prior information the children received on how it was brought about. In a similar experiment (Gelman and Bloom 2000), older children and adults saw an object but were given two diverging reports on how it came into being. In the unintentional version, a strip of cloth was accidentally caught in a machine, resulting in holes being punched at regular intervals. In the intentional account, a person took a pair of scissors and carefully cut evenly spaced holes. Participants were more prone to call the object a belt if they believed it was intentionally created.

The role of background information in our appraisal of design intentions is also detectable at the neural level. In an elegantly designed fMRI study (Steinbeis and Koelsch 2009), participants listened to a piece of electronic music. Half of them were told that the composition was written by a human composer, whereas the others were led to believe it was purely computer-generated. The first group of participants, but not the latter, exhibited a high activation in the anterior medial frontal cortex, superior temporal sulcus, and temporal poles, brain areas involved in the attribution of mental states and the inference of intentions of others. These experiments suggest that attributing design requires some background knowledge of the intentions of the maker: our judgment that something is an artifact depends on our knowing that the object was intentionally created. When natural theologians appeal to perceived design, they rely on background assumptions about the designer, that is, they assume theism. As John Henry Newman ([1870] 1973) revealingly put it, "I believe in design because I believe in God, not in a God because I see design."

Once we are familiar with specific classes of artifacts, we can reasonably infer that a particular member of a class was created with the intention of belonging to that class. When we see a broken chair in a pile of rubbish, we conclude that the object was created to fulfill a specific function (to sit on) and to be of a specific class (chairs). This stance also provides useful inferences when we have to identify classes of nonutilitarian objects, such as ships in bottles: although these boats will never sail, we still call them ships because the designer intended them to belong to this class of objects. It was on this inference that Paley relied in his assertion that the watch was a product of intentional design.

Conversely, knowledge of the designer's intention can help us to infer an object's intended function or identity. Take as an example Bloom and Markson's (1998) experiments in which three- and four-year-olds were shown featureless ovals that were purportedly drawn by a child with a broken arm, who, because of this, could not draw well. The young subjects were told that these were drawings of chickens (three vertical ovals) and a pig (one horizontal oval). When prompted, the preschoolers effortlessly identified the pig, because they reasonably inferred that the artist would draw objects from another category in a different way. Young children intuitively regard the creator of an artifact as having privileged knowledge about both its name and its intended function. This was illustrated by an experiment in which preschoolers saw objects that were given an anomalous label; for example, the experimenter showed the child a key-shaped object and said, "You are not going to believe this, but this is actually a spoon" (Jaswal 2006). Children were only willing to adopt the anomalous name if the experimenter referred to the object as something he had created, not as something he had merely found. Recognizing that the creator of an artifact has the prerogative to name it marks an important step in the development of the design stance.

The human propensity of inferring design may be due to the distinct evolutionary history of our species, which relies to a unique extent on tools. Whereas other primates mostly use unmodified objects as tools, archeological evidence for stone-knapping in hominids goes back as far as 2.5 million years ago (Semaw et al. 1997). By adopting the design stance, hominid children might have learned to use and fashion tools more efficiently. Indeed, comparative studies of social learning in children and chimpanzees (e.g., Horner and Whiten 2005) reveal stark contrasts in the way new tools are used: whereas children take the intention of the person who demonstrates these tools as guidance, chimpanzees rely more extensively on the physical properties of tools to figure out how they work. The design stance provides children with a useful heuristic to learn about their environment. It allows them to "ignore the actual (possibly messy) details of the physical constitution of an object, and [relying on] the assumption that it has a certain design, predict that it will behave as it is designed to behave under various circumstances" (Dennett 1987, 16–17). Without the design stance, we would not possibly learn to use and name hundreds of tools and other artifacts, but would perhaps be limited to the less than ten tool types a typical community of wild chimpanzees entertains (Whiten et al. 1999).

Intuitive Teleology

Whereas the design stance might have evolved in response to the regular use of artifacts throughout human evolution with the function of rapidly categorizing and using artifacts, humans also possess a natural propensity for teleological reasoning. As we saw in chapter 2, this inclination is most pronounced when thinking about biological entities, but it can apply to almost all categories of objects. Cross-cultural studies (e.g., H. C. Barrett 2004) indicate that humans have the intuition that animals and plants possess adaptations that are self-beneficial, such as claws for defense or thorns for protection against being eaten. Young children, however, not only attribute purpose to artifacts and biological adaptations, but also to entire organisms ("What are lions for?" "To go in the zoo") and nonliving natural kinds like clouds ("For raining")—a tendency Kelemen (2004) terms promiscuous teleology. Moreover, when given a choice between teleological and non-teleological explanations, preschoolers and elementary school children prefer teleological accounts. When asked whether rocks are pointy because of natural processes, such as "bits of stuff piled up for a long period of time," or because of teleological functions, for example, "so that animals could scratch on them when they got itchy," children typically endorse the latter (Kelemen 2003).

Around ten to twelve years of age, the preference for teleological explanations lessens, probably because adolescents acquire elaborate coherent mechanistic explanations through schooling. Although mountains can be climbed, few adults would claim that mountains are there to climb on. This is because our learned knowledge that mountains are formed by tectonic activity or volcanism is incompatible with teleological explanations, where the function provides a sufficient reason for why the structure exists. Remarkably, patients with Alzheimer's disease show a reemerging preference for teleological explanations. For example, they think that rain is there so that plants and animals can have water to drink and grow rather than the acquired explanation that rain occurs by water condensing into clouds and forming droplets (Lombrozo, Kelemen, and Zaitchik 2007). An increased tendency for teleology is also observed in adults with little schooling (Casler and Kelemen 2008). Formal education seems to reduce a preference for teleological explanations but cannot eradicate it. Indeed, when educated adults are forced to make very quick judgments, they too show a heightened acceptance of teleological explanations: when judging at a glance whether a statement is correct, they tend to endorse teleological, incorrect explanations such as "the Sun radiates heat because warmth nurtures life" (Kelemen and Rosset 2009). Reasoning strategies observed in children persist into adulthood but may be masked by secondary explanatory strategies. Once these become impaired (in the case of Alzheimer's patients) or are unavailable (in the case of speeded judgments or lack of education), the intuitive, evolved strategies of childhood reemerge. Although scientific education tends to lessen teleological reasoning, the tendency to apply teleology is not absent even among trained scientists. While their acceptance of incorrect teleological explanations is lower than in comparison groups, chemists, geoscientists, and physicists from top-ranked American universities are also more prone to endorse teleological explanations under speeded conditions than when they have unlimited time (Kelemen, Rottman, and Seston 2013). These experiments measure the extent to which people endorse spurious teleological explanations (such as "Trees produce oxygen so that animals can breathe") as a way to examine teleological tendencies. However, these experiments do not by themselves warrant the conclusion that teleological reasoning is mistaken, or that teleology applied to biological entities is always wrong. They only indicate that it is a pervasive reasoning strategy. We will now examine whether teleological reasoning in biology is incorrect.

Immanuel Kant ([1790] 1987, sec. 75) was still confident that there could never be a Newton of biology who could explain how a blade of grass comes about by reference to laws of nature that do not invoke intentions or goals. He thought teleology was indispensable to biology, as it brings order to our observations. However, even as Kant was writing, pre-Darwinian evolutionary theorists attempted to do precisely this—explain apparent design and teleology by reference to purely mechanical, mindless processes (see this vol., 75–76). By the twentieth century, philosophers were convinced that teleology or teleological function had no place in philosophy or scientific practice. Still, explanations in teleological terms remained (and still remain) pervasive in biology. As Ruse (1998, 16) observes, "Design language reigns triumphant in evolutionary biology." Although evolutionary biologists insist this language is metaphorical and just used as easy shorthand, it is clear that they cannot do without it, if only on a pragmatic level.

By the second half of the twentieth century, teleology became reintroduced in philosophical discourse, and there is continued debate on whether or not teleology has a place in evolutionary biology. Some philosophers of biology (e.g., Cummins 2002) have urged that teleology should be excised from biological discourse. Other philosophers and biologists (e.g., Ayala 1970) attempt to provide a naturalistic grounding for teleology, whereby the function of a particular trait can be regarded as a product of design through natural selection. Note that this does not require a conscious designer: the natural design is a consequence of the interactions of ancestral organisms that possessed this trait for their descendants' fitness. Authors using this principle are indeed quite adamant in their claim that biological function is a mind-independent, natural property of biological organisms "that does not involve the goals or purposes of a psychological agent" (Allen and Bekoff 1995, 611).

Are Humans Intuitive Theists?

Does the tendency to infer design in nature also entail an inference to a divine designer, as Paley, Śaṅkara, and others have suggested? At this point, developmental and experimental psychological data do not present a unified picture. Lombrozo et al. (2007) found that although Alzheimer's patients reasoned more teleologically, they nevertheless did not invoke God more frequently as an explanation than neurotypical older people. This experiment found no difference in religiosity between Alzheimer's patients and older people without this condition. In a study that probed Dutch primary school children's intuitive theories on the origin of species (Samarapungavan and Wiers 1997), the answers clustered together in different categories, including spontaneous generation, Lamarckism, and pure essentialism (i.e., animals and plants have always existed in their present form). Although many children made teleological inferences, only about 10 percent made explicit reference to God or intelligent design. On the other hand, a comparable investigation by Margaret Evans (2001) in the United

States found that a majority of ten-year-olds endorsed creationist accounts of the origin of species, regardless of their religious background. Kelemen and DiYanni (2005) obtained comparable results with British elementary school children, although the percentage of creationist accounts was significantly lower than with American subjects.

Several possible explanations might account for these findings. A strong position holds that humans are intuitive theists. In this view, creationism is a natural mode of reasoning that is only altered when children acquire explicitly nonreligious beliefs from their cultural environment. Bering (2006) defends this position, arguing that religious beliefs are biological adaptations that were directly selected for to enhance cooperation, altruism, and group cohesion. Kelemen (2004) argues that children are intuitive theists, as their intuitive teleology and design stance lead spontaneously to a belief in God. It is important here to note that the teleological stance that children have toward artifacts and natural items changes over time. Younger children, under the age of five, are more flexible about the telos of objects, whereas older children and adults think more rigidly about the functions of objects (functional fixedness, see chapter 2). The intuitions that drive the design argument are these mature, functional fixedness intuitions.

A weaker position (e.g., Bloom 2007) holds that religious belief itself is not innate but a byproduct of other cognitive adaptations such as agency detection and theory of mind. In this view, children acquire culturally transmitted religious beliefs easily because they key in on evolved propensities of the human mind. Here, the step from design to designer is not automatically made but needs to be made explicit, as Paley and others in fact did.

A third view holds that intuitive theism may be an innate cognitive capacity that depends on external cultural circumstances for its development. This position was proposed by Calvin, who argued that humans are endowed with an innate propensity to believe in God (*sensus divinitatis*) but that this propensity still requires cultural input for its development. Calvin ([1559] 1960, book 1, chapter 3) acknowledged that, depending on the environment where one is raised, the *sensus divinitatis* can result in a wide variety of religious beliefs. It seems to us that the empirical evidence surveyed here does not support the strong intuitive theism advocated by Bering and Kelemen. What is still required is an explicit link between design and a divine designer, a judgment that relies on intuitive probability.

Intuitive Probability: Can Chance Events Produce Order and Complexity?

Several fields of cognitive science, such as artificial intelligence, linguistics, and the cognitive psychology of reasoning, have experienced what Oaksford and Chater (2007) call a "probabilistic turn": when we cognize, we do not require or assume certainty (as when we make a deductive argument), but often rely on probabilistic reasoning. As the theologian Joseph Butler (1736) already remarked, we are finite beings restricted in time and space, and we cannot claim absolute knowledge but have to make do with probabilistic assumptions. Humans, like other animals, are sensitive to probabilities, and reason probabilistically in the face of uncertainty. Decades of research on perception, for instance, have shown that our detection of shapes depends on a variety of probabilistic inferences that our visual system automatically makes, based on shading, texture, boundaries, and shape recognition (e.g., de Gardelle and Summerfield 2011). Human infants rely on the probability with which syllables co-occur to distinguish words within continuous streams of speech sounds (Aslin, Saffran, and Newport 1998). Even animals with simple nervous systems, such as bumblebees, tune their behavior to probabilistic information, such as the likelihood that flowers of a particular type will contain nectar (Real 1991).

The design argument draws on an ability to assess posterior probability, the probability that is assigned after the relevant evidence is taken into account. How likely is it that the apparent design in nature was intentionally created or, alternatively, that it happened by chance? Early proponents of the design argument have taken the intuition that chance does not produce order as a starting point:

At this point must I not marvel that there should be anyone who can persuade himself that there are certain solid and indivisible particles of matter borne along by the force of gravity, and that the fortuitous collision of those particles produces this elaborate and beautiful world? I cannot understand why he who considers it possible for this to have occurred should not also think that, if a countless number of copies of the one-and-twenty letters of the alphabet, made of gold or what you will, were thrown together into some receptacle and then shaken out on to the ground, it would be possible that they should produce the *Annals* of Ennius, all ready for the reader. I doubt whether chance could possibly succeed in producing even a single verse! (Cicero [45 BCE] 1967, 213)

Cicero discarded the atomists' idea that chance collisions of elementary building blocks (atoms) formed the material world on the basis that chance has a low probability of producing order. Assuming that the twenty-one letters of the Roman alphabet are equally distributed into his "countless number," the chance of the first letter falling in the correct place is 1/21, the chance that the first two letters are correct is $1/21 \times 1/21 = 1/441$. The chance that the letters would produce the approximately 7,000 characters of the 600 lines that survive of Ennius's *Annales* (a now fragmentary epic poem on the history of Rome) is vanishingly small, being $1/21^{7,000}$. Cicero's intuition has been reiterated many times, including astronomer Fred Hoyle's image of hurling around scrap metal at random and happening to assemble a Boeing 747 ("Hoyle on Evolution" 1981).[1] Although all arrangements of the scrap metal are equally improbable, very few of them will fly; similarly, although all combinations of the twenty-one Roman letters are equally unique, very few of them will produce a legible text, let alone the *Annales*. William Dembski (1998, section 6.2) follows a similar line of thought in his *generic chance elimination argument*: one can infer that an event occurred by design if the probability of its occurrence is sufficiently small.

Although it is intuitively compelling, rejecting chance as an explanation for complexity and design is problematic because, as Sober (2002) notes, there is no probabilistic equivalent of modus tollens. In other words, we cannot state that

If hypothesis *H* were true, observation *O* would be highly improbable.

But *O*.

Therefore, *H* is not true.

The law of likelihood in statistics stipulates that it is not the absolute value of the probability of data under a single hypothesis that is to be considered but rather how the probability values compare under different hypotheses. The intuitive idea that improbability strengthens the existence of God is problematic in that it tacitly relies on an analogy between human and divine agency. When deciding whether human design or chance is responsible, we rely on empirical knowledge of what human agents in fact do. In an example from Kenneth Himma (2005), adapted from Dembski, suppose a political candidate's name appears first on the lists of voting ballots forty out of forty-one times. The probability of such an event occurring by chance is very small. But when we suspect that a county clerk rigged the list, we rely on tacit knowledge: that undecided voters are more likely to choose the first on the list and that the county clerk wants a particular candidate to win. Being an intelligent agent, it is not unlikely that he rigged the list. We also know of cases in which ballots were tampered with to win an election. Hence the hypothesis that the name was placed first forty out of forty-one times by design rather than mere chance becomes very plausible indeed.

In the case of divine action, by contrast, we do not have empirical knowledge to draw upon, and thus no assumptions can be made about what God would or would not do. While revealed theology can appeal to scripture to make assumptions about God's intentions, this route is not available for the natural theologian, whose knowledge about this does not appeal to sources outside of reason and empirical observation. Some authors, like Swinburne, appeal to inductions, made on the assumption of God's perfect goodness, to make predictions about what God would or would not do. For instance, assuming that God would perform the best action, he would create free agents like humans. However, such inferences rely on Swinburne's moral intuitions, which, as he admits, may not be reliable: "The moral intuitions on these matters that I am commending to my readers must inevitably be somewhat tentative, and, even if they are correct, very precise numerical values may not capture the resulting probabilities about what God will create" (Swinburne 2004, 123). Therefore, a natural theologian cannot straightforwardly appeal to God's intentions by using empirical evidence. Without the necessary background data to make the design argument an IBE, this argument relies on an analogy between human and divine agency. Again, this version of the design argument becomes an argument from analogy. As we have seen, this was successfully attacked by Hume, and it was precisely for this reason that Paley recast the argument from design into an argument to the best explanation, relying on an IBE.

The reliability of IBE depends on the amount and quality of the data and the relevance of the data to the conclusion. If insufficient evidence is available, it may well lead one to choose "the best of a bad lot" (van Fraassen 1989, 143). In fact, the bad-lot concern applies even if one has all the possible evidence, because one may simply have failed to conceive of the true theory with this evidence in hand. Cicero overlooked the principle of cumulative selective retention as an alternative to brute chance and intelligent design. According to this principle, the nonrandom retention of random events can, over time, result in complex, orderly features. An illustration of this is Dawkins's (1986) weasel program. The likelihood that a computer program that generates random combinations of letters will produce by pure chance a phrase from Shakespeare's *Hamlet*, such as METHINKSITISLIKEAWEASEL, is vanishingly small. However, allowing the computer to retain the correct letters at each attempt will produce the sentence in no more than $23 \times 26 = 598$ runs. Similarly, in Cicero's example, if each letter that falls in the correct place is selectively retained, we need at most $21 \times 7,000$ trials to complete what is now left of Ennius's *Annales*. When using an IBE strategy, most modern versions of the design argument do not take natural selection and its principle of cumulative selective retention into account as a viable explanation. Dembski holds that regularity, chance, and design exhaust the possibilities. He considers regularity and chance separately but fails to take the combination of chance and regularity into account, and it is precisely this combination that we know as natural selection (with random mutations and nonrandom, regular selective pressures). On their own they cannot generate the appearance of design, but combined, they can.

To be sure, in 1802, let alone in 45 BCE, natural selection was not in the pool of possible explanations. However, as Sander Gliboff (2000) demonstrates, Paley did have a range of alternative materialist explanations, of which we mention four. First, necessity: because everything has to have some form, it may as well be the present form; for example, the eye is the actual realization of one of the possible ways to fill an eye socket. Second, he considered infinite trial and error: given an infinite time and universe, every possible configuration of matter could be produced, some of which turned out to be viable life-forms that persisted and reproduced—an interesting precursor to the concept of natural selection, proposed by Epicurus, and described in detail by Lucretius ([ca. 50 BCE] 2007). Third, he discussed the claim that parts of organisms could arise before their function was determined, a forerunner of exaptation theory (Paley [1802] 2006, 38–41). These alternatives were being explored and hotly debated in Paley's time by early evolutionists like Buffon, Diderot, and d'Holbach. Paley seemed to have been familiar with these authors, as he mentioned Buffon explicitly and others implicitly. Fourth, he briefly discussed the special biological forces or organizing principles proposed by the so-called Göttingen school of German biologists, such as Blumenbach, Kielmeyer, and Reil (Paley [1802] 2006, 218–225). Although now rejected, the latter's Newtonian approach to biology that stipulated forces acting on biological entities, analogous to physical forces acting on physical entities, was conceivable and widely accepted at the time (Larson 1979).

Paley considered these alternative explanations carefully but ultimately dismissed them all, primarily for lack of evidence and lack of explanatory potential. For instance, he criticized Buffon's (1766) appeal to "internal moulds," which organized the organic particles that made up any individual creature (a sort of building plan) as a naturalistic explanation for why animals belonged to particular species. Paley thought this explanation was not an explanation at all—no more than the Greek essential forms explained design satisfactorily. By contrast, he thought that a designing mind was a satisfying explanation because it is "comprehensible to our understanding, and familiar to our experience" (Paley [1802] 2006, 224). Ultimately, Paley believed that even if these proposed naturalistic mechanisms and propensities are genuine, they still require a designer for their explanation. Biological forces and internal molds might explain the origin of biological design, but where did they originate? "I am unwilling to give to it the name of an *atheistic* scheme ... because, so far as I am able to understand it, the original propensities and the numberless varieties of them ... are, in the plan itself, attributed to the ordination and appointment of an intelligent and designing Creator" (Paley [1802] 2006, 224–225). This may strike one as a circular argument: he rejected these naturalistic mechanisms because he thought such mechanisms still require a designer, but that was the proposition that had to be proven. A more charitable interpretation is that Paley assigned a high prior probability to the existence of God, and hence to divine design, and a low prior probability to metaphysical naturalism. As will be argued in the next section, the likelihood of data can be meaningfully assessed only in relationship to hypotheses, which are accorded a prior probability.

A Rational Basis for Disagreement

If humans are prone to discern design and teleology in nature, why do some find the design argument more compelling than others? This may be due not to intrinsic differences in the way design and teleology are discerned but to differences in the prior probability people place on the existence of a divine designer. An interesting way to approach this problem is through an examination of how humans regard coincidences. For

Griffiths and Tenenbaum (2007), an event is a coincidence if it is judged to have a lower probability of occurring under our current theory of how the world works than under an alternative hypothesis. Coincidence plays an important epistemic role in scientific discovery: the geophysicist Alfred Wegener (1912) noted that the coastlines of West Africa and South America fit into each other like pieces of a jigsaw puzzle, that their geological strata match, and that the distribution of species on both sides of the Atlantic is highly correlated. He thought that this pattern was not a mere coincidence but that these continents were once joined and had drifted apart. The physician John Snow (1855) noted that cholera outbreaks in London tended to cluster at public water pumps and inferred that this was not a coincidence but provided evidence for his theory that cholera was transmitted through polluted water (rather than bad air, the then-favored theory). These examples suggest an intimate connection between coincidence and evidence. A coincidence occurs when the likelihood ratio in favor of an alternative theory is insufficient to overwhelm the prior odds against it. A coincidence becomes evidence when the likelihood ratio in favor of an alternative theory overcomes the prior odds against it and leads us to accept that alternative theory. Because people differ in the prior probabilities they assign to alternative hypotheses, what is a coincidence to one person can be compelling evidence to another.

In the case of the design argument, the competing hypotheses are H_{mat} (apparently purposive and complex structures arose strictly through natural, material causes) and H_{deo} (design as the result of a designer). In the framework of H_{mat}, the occurrence of ordered complexity and apparent design presents a coincidence. Given that chance events tend to produce disorder, the probability of this occurrence is extremely low. The theory of natural selection has successfully solved this dilemma because it relies on a combination of chance and lawlike processes. Indeed, no other naturalistic theory can explain why living things are improbably complex, why the interrelationships between their parts are highly functional, and why they exhibit features that enhance their probability of surviving and reproducing in their environment. Proponents of H_{mat} can find justification in the evolutionary explanation of design. To justify why they favor their view rather than H_{deo}, they can cite examples of maladaptedness and appeal to ontological parsimony because their explanation is restricted to physical processes. Under H_{deo}, the occurrence of design is not improbable because this theory explicitly proposes a designer who made the universe orderly and purposeful. Under these epistemic circumstances—but not under H_{mat}—design in nature becomes corroborative evidence for the existence of a divine creator. Next to this, natural theologians can also appeal to ontological parsimony because it reduces many kinds of explanation to one under H_{deo} (Swinburne 1968a).

This model of prior probabilities explains why evolutionary thinkers writing before 1859 did not accept the design argument. Even in Paley's time, not everyone was led to accept H_{deo}, although natural theological arguments were widespread and widely accepted. Early evolutionists such as Erasmus Darwin sought to describe biological forces that could assemble complexity in the same way that Isaac Newton had done for mechanics. These authors had a strong commitment to a materialist worldview, leading them to adopt the view that H_{deo} was unlikely even though they did not have a compelling causal explanation for the apparent design. The importance of prior probabilities may explain why the design argument, despite its intuitive appeal, failed to convince them. It may also explain the considerable cross-cultural variability in the acceptance of evolutionary theory. In one survey (Miller, Scott, and Okamoto 2006), about one-third of Americans believed evolutionary theory was absolutely false, whereas only 7 percent of Danes thought so. These differences correlate well with levels of religiosity in these countries, with Denmark having a lower regular church attendance rate (3 percent) than the United States (44 percent) (Verweij, Ester, and Nauta 1997).

For theists, design in nature provides compelling evidence for the existence of a creator. Take as an illustration the Thomistic tradition, which emphasizes the role of understanding and knowledge in belief. In this view, a successful natural theology would start out from self-evident premises, proceed by valid arguments, and reach the conclusion that there is a person such as God. As we have seen, humans are prone to discern design and teleology in nature. Within the epistemic context of H_{deo}, the perceived design in nature that is a universal feature of human cognition can be taken as a self-evident premise from which the existence of a creator can be argued. It is not a stand-alone argument that can convince those who do not believe in God, especially not since plausible naturalistic explanations have become available.

Is There Still a Place for the Design Argument?

Undeniably, the power of the design argument has been seriously weakened since Darwin and Wallace came up with natural selection as a naturalistic explanation of design. Given that the combination of random events and selective retention can explain most of the apparent design around us, can natural theologians still reasonably invoke the design argument?

Misrepresenting or altogether ignoring natural selection is the strategy most commonly adopted by proponents of the recent Intelligent Design (ID) movement. While it spans many fields (e.g., geology, biochemistry), its chief aim is to challenge Darwinian evolutionary theory. ID advocates that science points to irreducible complexity in nature that cannot be explained by naturalistic mechanisms (Behe 1996) and hence that such complex structures were intelligently designed. It differs from natural theology in that it relies exclusively on scientific, empirical research to reach this conclusion and does not equate, or at least not manifestly so, the intelligent designer with God. According to its proponents, ID does not require prior belief in God, as anyone who objectively looks at nature will realize that "a designer of remarkable talents is responsible for the physical world" (Dembski 1999, 111). ID nevertheless has a religious agenda: it wants to make religious belief a palatable, reasonable option for modern westerners; it aims to influence school curricula and ultimately wants to undermine the secularism and naturalism of modern science and society (Johnson 2000).

ID is not a very desirable position to take. Theologically, it employs a narrow notion of faith as the opposite of reason (Mongrain 2011). By insisting that religious belief should be exclusively based on reason, it goes much further than natural theology, which regards reason and faith as complementary, and natural theological arguments as faith seeking understanding. Ever since Augustine, theology (including natural theology) has been conceptualized as a form of seeking to understand what one already believes (we come back to this in the last section of chapter 5). To put it with Anselm of Canterbury ([11th c.] 2000, 93): "But I yearn to understand some measure of Your truth, which my heart believes and loves. For I do not seek to understand in order to believe, but I believe in order to understand. For I believe even this: that unless I believe, I shall not understand." This is the classical understanding of what Christian theology does, and ID is in diametrical opposition to this.

As a scientific research program, ID has not been able to establish itself
as a contender to evolutionary theory; unlike the latter, it has not produced substantive results. Its key concepts, *design* and *intelligent*, are not
sufficiently operationalized. For instance, the designer remains elusive, as
ID has scrapped all explicit references to theism.[2] However, without a specific referent (God), ID remains stuck in analogies to human design (Sarkar
2011). Dembski (1999) defines design in terms of *complex specified information*. Although formally sophisticated, this notion only operationalizes
complex as a measure of improbability, but does not operationalize *specified*
nor *information*. As we have shown, recognizing design requires background
assumptions about the nature and intentions of designers. Without such
background assumptions, the notion of design in ID will remain vacuous.
There is no serious attempt to define *intelligent* either, except by vague
analogy to human intelligence. For instance, it is unclear why irreducible
complexity, a term employed by Behe (1996) to denote the fragility and
interdependence of components in biological systems, should point to
intelligent design. If anything, should an intelligent designer not build in
some redundancy to make a system more robust and less prone to break-
down because of failure in a single component (Miller 2007, chapter 4)?
Human designers, for example, think of fire exits and emergency lighting
in buildings, spare parachutes and multiple motors in airplanes. As Sahotra
Sarkar (2011, 300) remarks, "All we have is a metaphor that works—to the
extent it does—because we all seem to share cultural ideas about what properties a supposedly intelligent theistic designer should have. We have no
positive specification of intelligence."

There is also a pragmatic reason not to align oneself with ID. Its aim to
help Christianity gain intellectual respectability through reason failed, as
its proponents are neither taken seriously in the scientific community nor
in mainstream natural theology. Augustine already warned against taking
positions that invite ridicule and decrease respectability.

Now it is quite disgraceful and disastrous, something to be on one's guard against at
all costs, that they [non-Christians] should ever hear Christians spouting what they
claim our Christian literature has to say on these topics, and talking such nonsense
that they can scarcely contain their laughter when they see them to be *toto caelo*,
as the saying goes, wide of the mark. And what is so vexing is not that misguided
people should be laughed at, as that our authors should be assumed by outsiders to
have held such views and, to the great detriment of those about whose salvation we

are so concerned, should be written off and consigned to the waste paper basket as so many ignoramuses. (Augustine [416] 2002, 186–187)

A more productive way for theologians and scientists to look at the argument from design is to treat it not as a scientific principle (as ID does) but as a metaphysical principle. Within this perspective, there are at least two cases in which a design position may still be defensible. A first case is presented by a position that endorses evolutionary biology but argues that God intervenes occasionally to fashion structures that could not have arisen through natural selection. One endorses intelligent design as a philosophical position but not as a scientific research program that conceptualizes evolution and design as competing scientific explanations. Theologically, it follows a distinction that is commonly made between God's general actions (which pertain to the universe as a whole and can be seen in the laws that govern physical, chemical, and biological processes) and special actions (which lie beyond normal physical processes). Whereas natural selection and other evolutionary processes belong to the former category, intervention in these or occasional design belongs to the latter.

This strategy is followed by Johnson and Potter (2005), who propose that human natural language may be the product of purposive creation. Adaptationist explanations require a plausible reason for why an adaptation evolved. Adaptations evolve in response to specific selective pressures and enhance the survival and reproduction of their bearers. For language, there are as yet no persuasive adaptationist explanations. We do not know what language is an adaptation for or how and when it evolved. Despite the proliferation of adaptationist accounts of the origin of language, such as social grooming, technological intelligence, cooperative hunting, and sexual selection, none of these hypotheses has been able to substantiate itself into a generally accepted theory. This leads Johnson and Potter (2005) to infer to the best explanation that purposive design brought language into being. Their position is distinct from ID in that they explicitly endorse evolutionary theory as the best explanation for complexity in the living world in general terms. One problem with this line of reasoning is that it relies on a God of the gaps argument. While God of the gaps arguments are not necessarily fallacious (as we will see in chapters 6 and 7), they have proven to be notably vulnerable in the past: once a convincing scientific account of the evolution of language is outlined (which is not unlikely, given the advances in our understanding of the genetic and developmental bases of language, its cultural evolution, and its emergence in human evolution), Johnson and Potter's argument may dissolve.

A second, perhaps stronger case is found in scientists and theologians who regard design and evolution as complementary rather than mutually exclusive explanatory frameworks. It avoids the vulnerability associated with God of the gaps arguments but relies on assumptions about how a divine designer would create the world. If the analogy with watches is maintained, such theistic evolutionary accounts could point out that watchmakers do not build watches from scratch but rely on the gradually accumulated innovations in timekeeping technology, which can be traced back to sundials and water clocks, to the introduction of the spring, to the modern digital watch. In fact, upon close scrutiny, very few inventions appear *de novo*; most are the result of a gradual and cumulative retention of favorable variations (Basalla 1988). For instance, the streamlined design of Polynesian canoes, which is close to optimum, can be traced through archeological and historical data as the gradual and unconscious retention of favorable variations, with the perilous ocean as the selecting agent (Rogers and Ehrlich 2008).

Since evolutionary processes play a pervasive role in human design, it is not an ad hoc move for the proponent of the design argument to regard evolution as a basic design feature of the created universe. The botanist Asa Gray (1888, 57) defended the view that natural selection is an "a-fortiori extension to the supposed case of a watch which sometimes produces better watches, and contrivances adapted to successive conditions, and so at length turns out a chronometer, a town clock, or a series of organisms of the same type." Theodosius Dobzhansky (1973, 127), one of the founding fathers of the modern synthesis, wrote, "The organic diversity becomes, however, reasonable and understandable if the Creator has created the living world not by caprice but by evolution propelled by natural selection. … Evolution is God's, or Nature's, method of Creation." The cell biologist Kenneth Miller argues that God has initiated natural selection and other natural evolutionary processes as an indirect way to create complexity and design. The undetermined nature of evolution through natural selection and other natural processes enabled the evolution of truly free, truly independent beings (Miller 2007, chapters 7 and 8). Theistic evolutionary accounts like these can invoke evolution, with its enormous creative potential, as a plausible mechanism for divine design.

Interestingly, both positions, materialism and theistic evolution, already existed in the earliest stages of evolutionary theory. Whereas Thomas Huxley did not provide room for God in his explanatory framework, Alfred Wallace and Asa Gray were theists who treated divine action as complementary with a scientific worldview, not as a competitor. Wallace (1870, 343), while continuing to endorse natural selection as the chief principle guiding the evolution of plants and animals, invoked intelligent design for the human mind: "The brain of prehistoric and of savage man seems to me to prove the existence of some power, distinct from that which has guided the development of the lower animals through their ever-varying forms of being." Wallace's theism, rather than a *volte-face*, was an integral part of his evolutionary thinking (Fichman 2001). Today, both schools of thought continue to exist side by side, with Richard Dawkins and Daniel Dennett as examples of strict materialists, and Kenneth Miller and Francis Collins as proponents of theistic evolution.

While theistic evolution may be an attractive option for the theist, it requires additional assumptions about the intentions of the designer. For example, Miller (2007) argues that God used a stochastic process because he wanted to create free beings that were genuinely distinct from their creator. The problem with this method of creation is that it seems to implicate God in an unacceptable amount of animal suffering that is a direct result of the winnowing process of natural selection. Animal suffering caused by predation, competition, and starvation is a byproduct of natural selection that seems difficult to reconcile with a benevolent creator (but see Southgate 2008 for a theological exploration and attempt to meet this challenge). While death and suffering were traditionally explained by appeal to a historical Fall of humankind, the current best scientific evidence is that these features of the world predate the evolution of humanity (see also De Smedt and De Cruz 2014). The atheist can reply that features of evolution, such as its generation of natural evil, undermine the design hypothesis. Some Christian theists have in turn responded to this by presenting more open, process-like theological conceptions, where God is not an aloof, external orchestrator, but became incarnate to suffer along with his creation. For example, John Polkinghorne writes:

In the lonely figure hanging in the darkness and dereliction of Calvary the Christian believes that he sees God opening his arms to embrace the bitterness of the strange world he has made. The God revealed in the vulnerability of the incarnation and the vulnerability of creation are one. He is the crucified God, whose paradoxical power is perfected in weakness, whose self-chosen symbol is the King reigning from the gallows. (Polkinghorne 1989, 68)

Discussions on the plausibility of theistic evolution indicate that theism and metaphysical naturalism cannot be purely evaluated by empirical evidence but rely on the prior probabilities their defenders assign to H_{mat} and H_{deo}. Indeed, the theist can argue that natural selection is not metaphysically incompatible with divine design, even if the mutations that are generated appear to be entirely stochastic from our perspective. The nontheist can argue that metaphysical naturalism is the most parsimonious explanation for natural selection, as it can explain features like suboptimal design and natural evil. A productive way forward for dialogue on design would therefore not be to focus exclusively on scientific evidence and its compatibility with the design hypothesis or naturalism, as proponents of ID or some atheist authors do. Rather, it would be to consider whether divine design through evolution provides an *overall* plausible worldview, or whether metaphysical naturalism is a more compelling framework.

Summary

This chapter has examined the cognitive basis of intuitions that drive arguments from design. These intuitions can be traced back to evolved inference mechanisms: the design stance, which leads us to treat complex and purposive structures as the product of design, and intuitive teleology, the propensity to discern purpose in nature. These inference mechanisms are universal, although they can be masked by formal education or strengthened by religious upbringing.

The step from design to designer is perhaps more explicit and relies on an argument to the best explanation. The plausibility of this argument relies on the prior probability one places on the existence of God. By making these differences in prior probability more explicit, theists (natural theologians, biologists, and philosophers) and materialist scientists and philosophers have a rational basis for disagreement. The reason why some find the design argument compelling and others do not lies not in any intrinsic differences in assessing design in nature but rather in the prior probability they assign to complexity being produced by chance events or by a creator.

5   The Cosmological Argument and Intuitions about Causality and Agency

The cosmological argument infers the existence of God from the existence of the universe. It has been developed in various traditions of natural theology (e.g., Christianity, Judaism, Islam, and Hinduism). Early examples include the Kalām (Islamic theological) cosmological argument, formulated by among others Ibn Rushd (Averroes) and al-Ghazāli, the second and third of Thomas Aquinas's five ways, Duns Scotus's argument from contingency, and cosmological arguments based on the principle of sufficient reason by Gottfried Leibniz and Samuel Clarke. Despite an equally distinguished list of critics, such as David Hume, Immanuel Kant, Bertrand Russell, and C. D. Broad, it still enjoys a widespread popularity in contemporary philosophy of religion (e.g., Craig 2003; Koons 1997; Swinburne 2004).

Our aim in this chapter is to investigate what cognitive factors underlie the intuitions that serve as premises for the cosmological argument. We propose that the enduring appeal of the cosmological argument is due at least in part to its concurrence with human cognitive predispositions, in particular, intuitions about causality and agency. Intuitions about causality underlie the inference to an external cause of the universe, whereas predispositions toward agency make God a natural candidate for this cause. Even modern versions of the cosmological argument that are couched in sophisticated formal terms are based on and ultimately stand or fall with the soundness of these intuitions. This chapter also examines the implications of cognitive science for the cogency of cosmological arguments.

The Cosmological Argument and Human Cognition

Cosmological arguments can be usefully categorized in three classes (see e.g., Craig 2003; Oppy 2009). The first, exemplified by Thomas Aquinas's second and third way ([13th c.] 2006, part 1, question 2, article 3, 5), relies on the observation that causes stand in relation to their effects as chains; as an infinite regress of causes is deemed impossible, this leads to the inference of an uncaused cause, that is, something that has itself as a sufficient cause. Aquinas's second way is the argument from first cause: as things cannot cause themselves, and as an infinite regress of causes is impossible, there must be a first cause that is itself uncaused. His third way observes that all natural things are contingent; in other words, they may as well not have existed. If everything were contingent, then even now nothing would exist, since things that do not exist only come into existence through things that already exist. Therefore, there must be something that exists of necessity, that is, it is impossible for this being not to exist. The second class, the Leibnizian cosmological argument, says that the totality of the world is a contingent being that requires a sufficient explanation for its existence (Leibniz [1714] 1898). The third class, exemplified by the Kalām cosmological argument, contends that all objects that have a temporal beginning must have a cause. Since the universe has a temporal beginning, it must have a transcendent cause. We will focus on this category.

The link between the cosmological argument and the structure of human reasoning was first proposed by Kant, who claimed that arguments from natural theology are unavoidable given the structure of human reason. The intuitions that underlie the cosmological argument, such as our propensity to look for explanations or to seek necessary causes, are regulative ideas of human reason, which bring "systematic unity into our cognition" (Kant [1781] 2005, A616/B644). The crucial difference between his and the present account is that Kant could only rely on introspection when considering causal intuition, whereas present-day philosophers can also draw on empirical data from cognitive science that, as we shall see, are very relevant for the debate. The case for the cognitive origins of the cosmological argument is better informed than it was more than two centuries ago.

Before proceeding, a caveat is in order. In chapter 1, we saw that there are two ways in which implications of CSR for natural theology can be examined. First, it can elucidate the causal history of natural theological intuitions. For the cosmological argument, these are causal intuitions. CSR can thus help to examine the origins of causal intuitions that underlie this argument. Second, it has implications for the justification of natural theological arguments in exploring how causal intuitions figure into the justification of these arguments and in considering whether it is appropriate to invoke them. Wilfred Sellars (1956) drew a distinction between the space of reasons and the space of causation. The space of reasons is the space of justification for beliefs: reasons (or candidate reasons) that present themselves to our judgment and that guide our inferences. The space of causes discusses the causal history of particular beliefs. Sellars held that these are two separate domains: the causes that underlie a particular belief do not, in themselves, say anything about their justification. What would constitute good reasons for accepting cosmological arguments? They need to be more than merely appealing—after all, as Davidson (1963) observed, there are many things that may hold a certain appeal but that we would not subscribe to. What can be a compelling reason for accepting the cosmological argument is that it concurs with basic human intuitions about causality and agency. To explain why we have these intuitions, we need to look at the space of causes: regularities in human cognitive development that shape causal intuitions in particular ways. When examining the cognitive origin of intuitions that underlie the cosmological argument, we are in effect proposing a causal factor that impinges upon the space of reasons—such a factor would not, strictly speaking, be a move within the space of reasons and, therefore, according to Sellars, would lack justification.

Nonetheless, from a methodological naturalistic perspective, there is no unbridgeable gap between causes and reasons, but a fundamental continuity in the causal natural order: human minds and the thoughts they form do not stand outside this natural order (Blackburn 2001). Consequently, naturalists have attempted to bridge the gap between causes and reasons. In many cases, the causal origin of a belief *does* say something about its justification. For example, meteorological observations lead to more justified beliefs about the future weather than consulting a crystal ball. And often, it is difficult to make a categorical distinction between causes and reasons. For instance, Susan Hurley (2003) has argued that evolutionary pressures cause animals to act in particular ways, for example, to look for food sources, seek out mates, or defend their territory. At the same time, animals are not robots: they can make flexible decisions within specific contexts that are driven by cognitive adaptations, for instance, in their social interactions with conspecifics. According to Hurley, we can say that these animals have reasons for their actions even though these reasons are never made explicit (by the animals themselves) and even though these reasoning processes are causally constrained by quite specific, ecologically relevant conditions.

If we accept that our cognitive capacities, like those of other animals, are causally constrained by evolutionary and environmental factors, how can we bridge the gap from causes to reasons? One influential proposal is due to John McDowell (1996), who has developed the notion of *Bildung* (upbringing and education). *Bildung* is the process by which we acquire habits of thought and action through experience, such as an ability to make ethical or normative judgments. To McDowell, these habits would be sufficient to constitute an individual's competence in the space of reasons. *Bildung* naturalism still requires an account of how these habits, which together constitute our rational capacity, are acquired. Such an account has not yet been outlined. Bill Pollard (2005, 76) has suggested that "developmental psychology could assist us in this endeavor." In chapter 2, we discussed the notion of core knowledge: developmental psychologists (e.g., Spelke and Kinzler 2007) have proposed that humans are equipped with a set of core principles that regulate their knowledge acquisition, among others in the domains of physics and psychology (e.g., agency detection). These principles are phylogenetically old, emerge early in development, and remain stable throughout adult life. According to developmental psychologists (e.g., Carey and Spelke 1996), core knowledge domains become elaborated and enriched through experience and education, but, crucially, they are not fundamentally revised. To put it differently: these maturationally natural capacities eventually are enriched with practiced natural skills, but they still constitute the basis of our thinking.

As will be argued further on in this chapter, intuitions about causality and agency that are present in young children are still regulative in the formulation of the cosmological argument. In other words, the habits that underlie our reasoning about causation are partly based on early-developed intuitions. If developmental psychologists are correct in proposing that core knowledge still "guides and shapes the mental lives of adults" (Spelke and Kinzler 2007, 89), these intuitions continue to play a role in shaping the reasoning that guides human inferences, including those in philosophical and theological reflection. Thus, our examination of the origins of the intuitions that underlie the cosmological argument can be situated within a moderate naturalistic framework that seeks to relate causes and reasons.

Causal Cognition and the Inference to a First Cause

Most cosmological arguments proceed in two steps: first, they establish that the existence of the universe must have a cause, and second, they identify this cause as God. In this first step, historical versions of the cosmological argument (e.g., the Kalām and Thomistic versions) rely on the causal principle: every contingent state of affairs has a cause of its existence.[1] To illustrate this, here is a formulation of the Kalām cosmological argument (see Shihadeh 2008 for an extensive treatment):

1. *Causal premise*: Whatever has a temporal origin (i.e., begins to exist) has a cause of its existence.

2. The world has a temporal origin (from the *ḥadīth*).

3. Therefore, the world must have an originator (from 1 and 2).

4. This originator must be eternal, otherwise it too must have an originator (from 1).

5. *Identification of God*: The originator is God.

Contemporary versions of the cosmological argument proceed in a similar way, although they attempt to replace intuitions with more rigorous logical argumentation. For example, Joshua Rasmussen (2010) provides an updated version of Duns Scotus's argument from contingency: all contingent, concrete objects can have a causal explanation. Using this causal principle, he derives the existence of a concrete, necessary being, wielding S5 modal logic (a system of logic that invokes notions like necessity and possibility): if it is possible for the obtaining of a concrete object, or its duplicate, to be causally explained, then it is possible for there to be a concrete object that is not contingent—in other words, necessary. According to S5 modal logic, if it is possible for a necessary object to exist, then that object necessarily exists, that is, it is not possible that it does not exist. While this result is impressive, it critically relies on the causal principle. To justify this principle, Rasmussen draws on his causal intuitions in his everyday experience of objects:

Consider, for example, your favorite armchair. Surely the armchair's existence *can* be the result of causal factors, such as a craftsman or factory machine piecing together materials. (Indeed, it certainly *was*.) ... The principle seems to apply to very small objects, too: neutrinos, for example, can be produced from proton collisions in a particle accelerator. It's natural to generalize: necessarily, any contingent concrete object can have a cause. (Rasmussen 2010, 185)

One proponent of natural theological arguments, C. Stephen Evans, acknowledges this role of non-inferential, intuitive beliefs in cosmological arguments:

> I want to suggest that what lies at the bottom of all or at least most forms of the cosmological argument is a certain experience of the world or objects in the world, in which they are perceived as mysterious or puzzling, crying out for some explanation. I shall call this experience that of "cosmic wonder." Even Swinburne's probabilistic argument, with its rigorous appeal to confirmation theory, rests in the end on such an experience. (Evans 2010, 60)

To proponents of the cosmological argument, the causal principle is a self-evident principle that hardly requires justification. William Lane Craig (2003, 117), for example, argues that it "seems obviously true—at the least, more so than its negation." Critics of the cosmological argument have called this assumption into question. Kant ([1781] 2005, A609/B637) contended that "the principle of causality has no significance at all and no mark of its use except in the world of sense; here [in the cosmological argument], however, it is supposed to serve precisely to get beyond the world of sense." Although we perceive the world in terms of causes, we cannot be sure that causes exist in the observer-independent world. At best, one can justify the causal principle by induction, for example, by arguing that it is constantly being confirmed in our experience, and that it holds a central place in modern scientific practice and in contemporary philosophy and theology.

For Hume (1779, 167), an explanation of the world in causal terms may be epistemically satisfying, but such an explanation is only "an arbitrary act of the mind, and has no influence on the nature of things." By this, Hume (1748, essay 4, part 1) meant that the causal principle is not metaphysically necessary but, rather, that it results from a psychological disposition where cause and effect are joined in the mind of the observer. The observer relies on experience, in particular on a constant conjunction of events. Although Hume's ideas on causal cognition are no longer followed in cognitive science, his assertion that our detection of causes requires assumptions on the part of the perceiver and that such causal reasoning lies at the basis of the cosmological argument is still sound.[2] As will be shown in more detail below, our causal intuitions in everyday domains closely match those employed by the cosmological argument.

From an early age onward, humans seek causal explanations to account for everyday observations, such as why water expands upon freezing or why a relationship has failed. We have an intuitive feel for whether or not an explanation is satisfying. Satisfying explanations often invoke generative causes. People tend to infer causes spontaneously, without conscious deliberation and in the absence of instructions to do so (Hassin, Bargh, and Uleman 2002). As we saw in chapter 2, our intuitive ontologies rely on principles that invoke unobservable causes, such as mental states, physical forces, and hidden essences. In the domain of intuitive psychology, preschoolers invoke unobservable mental states to explain the behavior of agents: they spontaneously attribute beliefs, desires, and intentions to them and realize that mental states can differ from the actual state of the world. For intuitive physics, young children posit unobservable physical forces and properties to account for the motion of inanimate objects. They have, for instance, the intuition that unsupported objects fall downward due to gravity, and that one object can set another in motion when in direct contact. In the domain of intuitive biology, young children from about three years old posit invisible biological properties to account for the growth and behavior of biological organisms. Such hidden properties are invoked to explain why apple seeds planted in a flowerpot will still grow to be apple trees, or why caterpillars turn into butterflies. Whereas Hume thought that causal cognition relied on a superficial detection between the conjunction of events, young children and laypeople posit deep, structural causal features that relate to their effects. Indeed, it turns out that both children and adults prefer explanations that appeal to unobservable, hidden features over explanations that simply state that two events were merely associated in space and time (Shultz 1982; Ahn et al. 1995).

The ability to figure out causes has also been demonstrated in nonhuman animals, such as apes, which can use causal cues to find the location of hidden food (Bräuer et al. 2006). Next to this, human children of four years and older as well as adults are able to infer causes of events that they never experienced before, involving objects that they are totally unfamiliar with. In a classic series of experiments, Thomas Shultz (1982) showed Malinese children from a horticultural society, who were unfamiliar with Western technology, a variety of causal events involving flashlights and tuning forks. In these experiments, the participants were more likely to say that a tuning fork that was struck caused a box to resonate rather than a tuning fork that was closer to the box but that was not struck. Similarly, Western children and adults preferred a generative account of causality to explain why the propeller of a Crookes radiometer (an instrument for measuring electromagnetic radiation that none of the subjects had ever seen) began to spin when a flashlight was turned on.

Cosmological arguments postulate a cause for the existence of the universe, a unique state of affairs that cannot be compared to other events. From a psychological point of view at least, such a view is not problematic or unintelligible, as humans spontaneously make causal inferences about events with which they had no prior experience. To explain the actions of other agents, young children spontaneously appeal to invisible causes in the psychological domain, first desires and intentions (around the age of two), and later also beliefs (between the ages of three and five). Remarkably, children of this age do not understand that behaviors by agents can be accidental or caused by factors other than the agent's intentions and desires. Until the age of four, children tend to believe that sneezing and other involuntary biological actions are done on purpose and that knee-jerk reflexes are as intentional as voluntary leg kicks (Rottman and Kelemen 2012). Blindfolded three- and four-year-olds whose hand is guided by an experimenter to produce a drawing will insist that they intended to create that drawing, appealing to their own mental states as causes even if their hand movements were purely passive (Montgomery and Lightner 2004). This suggests that in the psychological domain, the default explanatory stance is to appeal to unobservable mental causes.

As these examples from intuitive physics and intuitive psychology indicate, humans routinely posit unobservables as underlying causes. Can other animals do the same? One set of experiments that compared the behavior of human children and adult chimpanzees indicates that preschoolers, but not chimpanzees, attempt to seek a cause for their failure to perform a task (Povinelli and Dunphy-Lelii 2001). In this study, the participants were taught to place an oblong L-shaped block in an upright position. When the block had been visibly tampered with so as to make the task impossible, both preschoolers and chimpanzees examined it extensively. However, when the block showed no external signs of manipulation, only the children explored it from different angles to attempt to find a reason why the task could have failed. The apparent inability of nonhuman animals to attribute invisible causes has been demonstrated in other domains as well.

While nonhuman apes such as chimpanzees have a sophisticated theory of mind, they seem to be exclusively focused on the observable elements of social cognition, such as eye direction. As we've seen, humans (including seven-month-old infants) automatically represent the beliefs of others: their expectations of an object's location are influenced by another agent's belief, even if it is irrelevant to the task. By contrast, rhesus monkeys are not influenced by another agent's belief in their expectations of a ball's location (Martin and Santos 2014). It is thus perhaps not surprising that whereas human children pass verbal false belief tasks between the ages of three and five and nonverbal false belief tasks from about thirteen months onward (Surian, Caldi, and Sperber 2007), no nonhuman ape has yet passed the false belief task (Call and Tomasello 2008). To explain these findings and similar results in other domains, Vonk and Povinelli (2006) propose that humans may be unique in their ability to conceptualize unobservables such as God, ghosts, gravity, and other minds. While validating this claim requires more systematic comparisons between humans and other apes in a variety of tasks, it is indeed remarkable that human children are able to infer unobservables and that they use these inferences in so many different domains. The early emergence of this ability and its sometimes inappropriate use by children suggest that looking for unobservable causes is maturationally natural for humans. The attribution of an unobservable cause for the universe in the cosmological argument is made possible by this universal human cognitive disposition to readily infer unobservable causes.

Most proponents of the cosmological argument also argue for the necessity of a cause for the existence of the universe. This part of the cosmological argument may also be informed by maturationally natural cognitive predispositions. Humans seem to be more prone to infer deterministic causes (causes that always produce their effects) than stochastic causes (causes that sometimes produce their effects). To test whether children prefer deterministic over stochastic causes, Schulz and Sommerville (2006) presented preschoolers with a lamp and a remote control with a switch. They were shown that sliding the switch turned the light on or off. They also witnessed that without a ring on top of the box with the lamp, the lamp would not be switched on. Participants were assigned either to a deterministic condition (the lamp always works if the switch is flipped), or a stochastic condition (the switch only sometimes turns the lamp on). After showing the event several times, the experimenter revealed a flashlight, which was concealed in her hand. Children from both conditions were asked to ensure that the lamp would not work. They could choose to either remove the ring or take the flashlight. Almost all children from the deterministic condition chose the ring. However, most children from the stochastic condition chose the flashlight. Schulz and Sommerville (2006) interpret these results as indicating that preschoolers, when faced with a stochastic causal pattern, will nevertheless try to find a deterministic cause to explain the apparently stochastic behavior of the lamp. After all, the participants from the stochastic condition also saw that the ring prevented the lamp from being turned on. So, interestingly, when they were given the choice between an observable cause and an unobservable cause that might explain the stochastic behavior deterministically, they preferred the latter, deterministic cause. They are, in the words of Schulz and Sommerville (2006, 432), "causal determinists."

This bias toward deterministic causal factors persists into adulthood. For example, Kathleen Metz (1998) compared the ability to infer stochastic causes of physical events in kindergartners, school-age children, and adults. She found that the ability to recognize stochastic causes increased with age, probably an effect of education. Nevertheless, like the children, the majority of adult participants continued to infer deterministic causes for some stochastically caused events. It seems that the causal reasoning that lies at the basis of the cosmological argument is not an arbitrary act of the mind, but rather a way of reasoning that is both obvious and intuitive to humans: we readily infer generative causes for events, we routinely deal with unique states of affairs, we habitually infer unobservable causal mechanisms, and we have a preference for deterministic causes.

Intuitions about Agency in the Identification of God

The second step of most cosmological arguments consists of an identification of the necessary external cause of the universe with God. This second step is important because, as Hume (1779, 164–165) already observed, one could simply argue that the material universe is metaphysically necessary. Moreover, even if we grant that the universe has an external cause, what reason do we have to identify that cause with the God of traditional theism, an infinitely powerful, all-knowing, eternal, and perfectly good being? It seems reasonable to suppose that if there is an external cause to the universe, it must be a powerful entity. This does not mean that it is a person, let alone God. To justify the identification of God, Craig (2003) proposes an argument of the following form:

6. The cause of the universe is timeless and immaterial.

7. The only entities of which we know that can be timeless and immaterial are minds or abstract objects.

8. Abstract objects cannot cause something to come into existence.

9. Therefore, the cause of the universe is a mind (this could be further specified as a timeless, immaterial mind).

As William Rowe (2005, 114–115) observes, this argument as it stands is invalid. In order to be valid, the conclusion should be:

9a. The only entity we know of which can be the cause of the universe is a mind.

Even in this revised form, the argument relies to an important extent on our finite, human intuitions about causality, where a person is regarded as the cause for any occurrence—in this case, the universe.

Craig's (2003) characterization of minds as timeless and immaterial entities is akin to intuitive psychology. As we saw in chapter 2, humans make an intuitive distinction between physical and psychological causes. The extent to which children are radical intuitive dualists remains a matter of debate (e.g., Bloom 2004; Hodge 2008), but at a minimum, humans intuitively expect agents to behave differently from non-agents; for instance, agents can perform actions without first having an external cause act upon them. This basic intuition still plays a role in the natural theological writings of contemporary philosophers of religion. For example, Swinburne distinguishes two kinds of causes (physical and personal), which require two types of explanation, scientific and personal. To explain the existence of the universe as a whole, one cannot posit physical causes since there are no physical causes except for the universe itself and its parts. Therefore, the universe is either a brute inexplicable fact or it is explained in personal terms. Swinburne (2004, 142–145) argues in favor of the personal explanation: he invokes a person, God, who freely chooses to create and sustain the universe. Given that the physical universe is extremely complex, whereas God is simple (in the sense of undivided, not composed of parts, etc.), Swinburne (2004, 147) argues that the theistic explanation is the more likely, as it is the more parsimonious: "The need for further explanation ends when we postulate one being who is the cause of the existence of all others, and the simplest conceivable such—I urge—is God."

The identification of God as the necessary cause of the universe can be traced back to human intuitions about agents as causes. Swinburne's distinction between physical and personal causes has parallels in the cognitive psychological literature: humans draw an intuitive distinction between events that are caused by purely physical processes and those caused by agents (Gelman and Gottfried 1996).[3] One of these distinctions that arises already in infancy is that agents, but not inanimate objects, are able to influence the behavior of objects from a distance (Spelke et al. 1995). Preverbal infants seem to appreciate that only agents can create order: they exhibit surprise (as measured by a longer looking time) when a rolling ball apparently causes a disorderly heap of blocks to become neatly stacked, but not when an unseen agent (hidden behind a screen) performs the same thing (Newman et al. 2010). This intuitive distinction between objects and agents as two types of causes is a core principle of human reasoning that persists into adulthood. Neuroimaging studies (e.g., Martin and Weisberg 2003) indicate that the perceptions of mechanical and agent-based motions are subserved by distinct and largely nonoverlapping brain areas. In particular, only motions performed by intentional agents activate areas reliably involved in the attribution of mental states to others, for example, the superior temporal sulcus, involved in intuitive psychology and agency detection, versus the posterior temporal lobe, subserving the manipulation of tools. The universe exhibits a high degree of order. Our attribution of its origin to an intentional agent is furthered by these stable intuitions about agents as causes.

Purposiveness is a decisive cue to favor agency. When adults watch simple geometric objects moving about on a screen, they interpret those motions as agent-like and explain them by reference to internal mental states if the objects appear to move in a goal-directed manner (Scholl and Tremoulet 2000). Experimental studies suggest that the ability to identify an agent as the cause of an event arises early in development. Twelve-month-olds witnessed a beanbag landing on a stage; subsequently an object appeared on stage that could be interpreted as the cause of this event (Saxe, Tenenbaum, and Carey 2005). The infants looked least long when a human hand appeared and significantly longer when a toy train or plush animal was shown, indicating that they expected the hand but not the toys to be the cause of the event. This implies that infants assume an agent to be the cause of a contingent event. This preference for agents was also shown in other studies. Gelman and Gottfried (1996) showed preschoolers different kinds of objects (animals, wind-up toys, other artifacts) that, under some conditions, moved without any apparent external cause. In the case of the animals, the children mostly referred to internal, biological features. By contrast, for the artifacts, they were much more likely to attribute the motion to a person. They expressed their surprise when they saw artifacts moving by themselves and frequently appealed to invisible agents, for example, "I think another person invisible [sic] did that again" (Gelman and Gottfried 1996, 1980). Children appeal to agents as causes when they have to explain the origin of artifacts and even of plants, animals, and natural kinds. As we saw in the previous chapter, this led Kelemen (2004) to call young children intuitive theists because they discern design and purpose in nature, regardless of whether or not they were brought up in a religious environment.

Explanations that postulate agents as causes of contingent events are intuitively appealing and epistemically satisfying. Humans have the intuition that beliefs, desires, and intentions directly cause actions. They regard them as generative causes that bring about actions (e.g., one forms the desire to lift one's hand, and the hand is lifted), although the actual cognitive processes that underlie human actions are far more complicated. As the social psychologist Daniel Wegner (2003) observes, personal accounts do not require further causal explanations. The intentions of an agent seem to us, intuitively, good ultimate explanations that require no further exposition. So when asking why John lifts his hand, the answer "because he wants to" provides a plausible account and does not prompt a search for further elucidation.

By contrast, an infinite regress of causes is not in line with our untutored ways of explanation. Experimental studies by Frank Keil (2003) indicate that folk explanatory schemas are typically very shallow: once a cause is identified, there is no further quest for causes. The amount of detail and sophistication within such explanations is typically also very coarse: laypeople, for instance, overestimate the extent to which they know how a helicopter or a zipper works. It is thus unsurprising that infinite regresses of causes are highly counterintuitive. Many versions of the cosmological argument, such as Aquinas's second way, which postulates a first cause, assume from the outset that an infinite regress is impossible. From a psychological perspective, postulating agents as causes is more intuitively compelling than postulating either uncaused physical events or infinite regresses of physical causes.

Although the step to an agent is intuitive, the step to a single omniscient and omnipotent creator seems somewhat far-fetched from a cognitive psychological point of view. Why might proponents of the cosmological argument favor God? Are they mainly motivated by background beliefs that are specific to their culture, that is, because natural theologians and philosophers of religion who have developed cosmological arguments stem from theistic religions? Our alternative explanation is that all-knowing and very powerful agents are intuitively more appealing than normal agents. As we saw in chapter 3, human reflective reasoning about other minds favors positing superhuman properties such as omniscience and omnipotence. It is easier to think that an agent's mental states correspond to the world (reality bias) than to attribute various cognitive and physical limitations to her. As we've seen, in reflective reasoning about other minds toddlers start out attributing omniscience to all agents (e.g., their parents), but later in cognitive development they restrict this to God. Children come to realize through experience that their parents do not know everything, but as they constantly receive testimony that God is omniscient, they have no reason to scale down that expectation. From a cognitive point of view, the inference to an omniscient creator of the universe is not an unlikely step.

How do early-developed intuitions about causality and agency figure as reasons for holding the cosmological argument? There are two ways in which their role for developing or endorsing the cosmological argument can be articulated. The first, which an agent can do introspectively, corresponds most to the commonsense notion of reason. It consists of a reconstruction of a reasoner's subconscious states when she was formulating the argument. Such states could include, for example, habitual modes of making inferences that cohere with her intuitive ontologies, in particular, her beliefs about how agents can be causes of physical events. In this first strategy, a defender of the cosmological argument might reason that "the appeal to an agent as first cause that underlies the cosmological argument coheres well with my intuitions about agents and causes in general. Thus, the cosmological argument is based on reasonable assumptions." Such appeal to intuitions is very common in philosophy. Indeed, philosophers frequently use their own phenomenological feeling of whether or not something is intuitively plausible as evidence for particular philosophical theories. Intuitions often strike them as compelling or even as obviously true. If philosophers do this frequently in areas such as epistemology, philosophy of language, and ethics, it seems reasonable to presume that they likewise appeal to the prima facie plausibility of intuitions in the domain of philosophy of religion.

A second conception of the relationship between intuitions and accepting arguments does not rely on introspection but on the assessment of others. This view stems from the worry expressed by experimental philosophers that our intuitions are not always reliable (e.g., Nichols, Stich, and Weinberg 2003). It has been suggested that methods other than introspection, such as cognitive psychology, are more useful to assess the role intuitions play in our evaluation of philosophical arguments. Developmental psychologists can formulate informed ideas about how the mind draws certain conclusions and what cognitive stable intuitions may underlie these. As outlined earlier, many developmental psychologists (e.g., Carey and Spelke 1996) subscribe to the view that early-developing principles of human reasoning are not fundamentally revised over time. Rather, they become enriched and sophisticated through experience and education. Obviously, the causal inferences that scholars trained in philosophy and theology formulate are far beyond the rudimentary causal understanding of young children. Yet, if these developmental psychologists are correct, there are good reasons to assume that philosophers and theologians are still guided by these early developed, cognitively stable intuitions about causality and agency. These intuitions can be explicitly articulated, addressed, and challenged in philosophical and theological discourse as, for example, in the discussion between Grünbaum (2000) and Craig (2001) on intuitions that underlie the cosmological argument. The fact that philosophers can do this is compatible with our claim that these intuitions have their origin in normal cognitive processes. As Timothy Williamson (2007, 3) remarks, "Neither their content [of intuitions in philosophy] nor the cognitive basis on which they are made need be distinctively philosophical." Instead, one should not be surprised that reasoning skills that are used in philosophy and theology are what Williamson (2007, 136) terms "cases of general cognitive capacities used in ordinary life, perhaps trained, developed, and systematically applied in various special ways, just as the cognitive capacities that we use in mathematics and natural science are rooted in more primitive cognitive capacities to perceive, imagine, correlate, reason, discuss." In the case of causal cognition, we have indicated continuities between commonsense causal reasoning and causal inferences that underlie the cosmological argument, such as the inference to causes of unique events, the preference for a necessary cause, and the favored postulation of an agent as cause.

Evolutionary Debunking Arguments

Given that intuitions about causality and agency play an important role in the formulation and acceptance of the cosmological argument, what are the implications for its cogency? In what follows, we consider implications for the justification of the cosmological argument from externalist and internalist perspectives. In epistemology, there is an enduring debate on what makes beliefs justified. According to externalists, beliefs are justified because of factors external to the person, in particular, the way her beliefs relate to the external world; she need not be aware of these. According to internalists, beliefs are justified because of factors internal to a person, such as reasons that become available through introspection. While she need not be consciously aware of them, they must in principle be accessible.

According to one externalist view (e.g., Dennett 2006), cognitive accounts of religion cast doubt on the reasonableness of religious beliefs. As we will see in more detail in chapter 9, this can be placed within the broader context of evolutionary debunking arguments: evolution can result in cognitive faculties that produce false beliefs. In some cases, cognitive processes might deviate from the truth due to a fitness trade-off between accuracy and efficiency. Given that animals have limited time and resources, they will sometimes be better off with fast heuristics than with faculties that are slow and always truth-preserving. In this view, the propensity to attribute causes may be a useful heuristic rather than an accurate reflection of the structure of the world. Sometimes natural selection will promote cognitive faculties that err on the side of safety, especially when one has little information and when the costs of false positives and false negatives are asymmetric. This asymmetry can lead to the evolution of cognitive predispositions that are triggered easily and that give rise to many false positives (Guthrie 1993; Stephens 2001). Arguably this is the case for causality and the inference of agency. Suppose one hears a noise in the night: assuming that the sound is uncaused when in fact it is caused by an intruder, a false negative is potentially far more costly (being murdered in one's bed) than the cost of a false positive, where one investigates in vain and goes back to sleep. Perhaps the inference of agential causation in the cosmological argument is an instance of a false positive.

Other authors maintain that our cognitive faculties work reliably under normal circumstances but not in situations that are remote from the conditions in which they evolved. Accordingly, the evolutionary origin of causal intuitions provides prima facie evidence against the soundness of the cosmological argument, as it applies causal intuitions to a domain that is not ecologically relevant, namely, the universe as a whole rather than states of affairs in our everyday life. Stewart-Williams presents the following evolutionary debunking argument:

We should be extremely cautious about accepting that there must be a causal answer to the question of why there is something rather than nothing. One popular answer to this question is to posit God as First Cause. However, we cannot rely on the intuition that there must be an ultimate cause for the universe as a whole. Thus, an important philosophical implication of evolutionary psychology is that it weakens the First Cause argument for the existence of God. (Stewart-Williams 2005, 801)

This type of criticism is ancient, predating the emergence of evolutionary psychology, as it already occurs in seventh-century responses to Hindu cosmological arguments. For example, the Buddhist atomist Dharmakīrti argued that cosmological arguments violate the boundaries of legitimate inductive extrapolation, as they are grounded in our ordinary experience of causes and effects in everyday circumstances. One cannot generalize from common experiences to a unique being such as God (Dasti 2011). A problem with this line of reasoning is that we simply do not know whether the attribution of a cause for the existence of the universe is off-track. Our causal intuitions are often correct (e.g., the noise in the night was caused by the cat), and there is no a priori reason to assume that they are off the mark when we apply them to the universe as a whole. Arguing that causal cognition does not work in the case of theism begs the question; in other words, it already assumes that there is no God.

Should we doubt either the causal principle itself or its applications outside the domain of commonsense reasoning, we would be faced with the unwelcome consequence that causal cognition is unreliable in the domain of science. After all, science has only developed in the last few hundred years and therefore does not constitute an ecologically relevant domain for our cognitive faculties. Moreover, holding erroneous scientific beliefs (such as the Earth is some 6,000 years old) does not seem to have a negative impact on human fitness as fundamentalist families tend to be large. The quest for causes remains an important, even essential, part of scientific practice, where it is supplemented with a sound methodology that yields results that if required can be replicated. The continuity in causal cognition between young children and adults has led some psychologists (e.g., Brewer, Chinn, and Samarapungavan 2000) to liken children to scientists, engaged in theory formation and hypothesis testing. Others have reversed this analogy, arguing that in fact scientists are like children: Gopnik and Meltzoff (1997), for instance, regard science as a byproduct of our universal search for causal explanations, emerging in early childhood. Should scientists still appeal to causes? Even those philosophers of science (e.g., Norton 2007) who argue that cause and effect are not fundamental concepts of science and that science is not fundamentally governed by the principle of causality concede that invoking causes remains an indispensable heuristic of actual scientific practice. This prominence of causal cognition illustrates the continuity between everyday and scientific reasoning. It therefore seems that rejecting the causal principle outside of the domain of commonsense reasoning comes at a high price. We will return to this *problem of collateral damage* in chapter 9.

Internalist Justification

The cognitive science literature on causal cognition can also throw light on internalist justification. The majority of our beliefs are held non-reflectively. Non-reflective beliefs can arise, for example, through perception (e.g., there is a red car parked in the street), or as a result of intuitive ontological knowledge (e.g., this unsupported pen will fall downward). Sometimes we reflect on these beliefs and other mental states and on our reasons for holding them. In this way, we can come to reflectively endorse our beliefs (Kornblith 2010) or, alternatively, decide that the non-reflective belief in question is not justified and reject it. Suppose someone is already a theist or atheist without deliberate reflection because she grew up in a theistic or atheistic environment. Natural theological argumentation, such as the cosmological argument, can provide reflective reasons for why one's original belief is justified or unjustified. This reflective endorsement is thus an important psychological process that underlies internalist justification. Recall that internalists hold that beliefs are justified because of factors internal to a person. We will here review how the assessment of the cosmological argument relates to the overall worldview of philosophers and then move on to the question of what role it can play in internalist justification.

In spite of their appeal to intuitions that humans widely share, natural theological arguments do not seem to be universally convincing. They may provide those who already believe the conclusion (e.g., there is a God) with additional rational grounds for belief, but they usually fail to persuade those who are skeptical. Kelly Clark (1990, 41–46) contends that natural theological arguments are not classical proofs—sound arguments that will convince everyone of their conclusions—but *person-relative proofs*, which provide evidence for beliefs one already holds. Jennifer Faust (2008) proposes that natural theological arguments are a form of faith seeking understanding—a way for theologians or atheologians to rationally support what they already believe on faith. If this is correct, natural theological arguments can play an important role in internalist justifications for theism or atheism. A crucial element in Faust's argument is that religious arguments "beg the doxastic question," that is, their assessment is influenced by the prior belief or disbelief in their conclusion. These arguments are not, strictly speaking, circular since they do not state their conclusion in their premises, but the fact that they require belief in the conclusion in order to be persuasive makes them question-begging in a doxastic (i.e., belief-related) sense. For example, many instances of suffering seem gratuitous, which means they do not seem to serve a greater good and could easily have been prevented by an all-powerful God. This observation is a key premise of the evidential argument from evil. Rowe (1979), however, submits that someone who believes there is an omniscient, omnipotent, wholly good being may simply not accept this premise; that is, she might deny that there is gratuitous evil.

One of the authors (Helen) empirically tested whether natural theological arguments are indeed person-relative. If they are, theists should rate higher the arguments in favor of God's existence, and atheists should rate stronger the arguments in favor of atheism. She conducted an Internet survey among 802 philosophers to ask how strong they find natural theological arguments, rating them from very weak to very strong.[4] As expected, religious belief/unbelief strongly influences the assessment of natural theological arguments. Helen computed a combined score of all arguments for theism and all arguments against theism. Since each argument could be rated from 1 to 5, and there are 8 arguments for and 8 arguments against theism, the minimum combined score is 8 (i.e., all arguments are rated as very weak), and the maximum combined score is 40 (i.e., all arguments are rated as very strong). She found that theists rate natural theological arguments that support theism more positively than atheists (mean of 25.5 for theists versus mean of 13.4 for atheists). Conversely, atheists rate arguments that undermine theism more strongly than theists (mean of 25.6 for atheists, mean of 17.7 for theists). Agnostics occupy an intermediate position (mean of 16.6 for arguments in favor of theism and 21.5 for arguments against theism). These differences were statistically significant (De Cruz in press).[5]

For the cosmological argument, the median response for theists is 4 (i.e., they rate the argument as strong), whereas both atheists and agnostics have a median of 2 (i.e., weak). This is a statistically significant result.[6] Figure 5.1 shows the responses in three box plots. Philosophical specialization also has a positive effect on the assessment of the cosmological argument, but this was more modest than the effect of religious orientation. Controlling for religious belief, philosophers of religion rate the cosmological argument stronger than those who do not have this specialization.[7]

This preliminary study suggests that religious belief indeed influences how strong one finds the cosmological argument, a finding that is in line with discussions between philosophers of religion. Take the Kalām cosmological argument. Theists tend to find its premises (whatever begins to exist has a cause of its existence; the world began to exist) more plausible than nontheists. Proponents often appeal to the big bang theory as scientific support for these premises. For example, Craig (2001, 373) writes, "Contemporary interest in the argument arises largely out of the startling empirical evidence of astrophysical cosmology for a beginning of space and time. On the standard Big Bang model the universe originates *ex nihilo* in the sense that at the initial singularity it is true that *There is no earlier space-time point* or it is false that *Something existed prior to the singularity.*" This is correct, but the big bang theory describes only the expansion and cooling of the universe and says nothing about its origin. Within the standard interpretation of this theory, the universe inflated from a tiny speck to the size it is today and will continue to do so. Given that matter cannot be infinitely packed, we cannot trace the universe back to an infinitely packed state. This seems

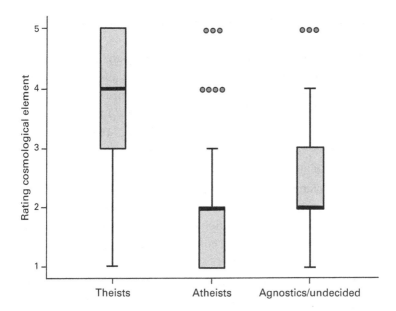

Figure 5.1
Evaluations of the cosmological argument by theists, atheists, and agnostics. Participants were asked to rate the cosmological argument from very weak (1) to very strong
(5). The thick horizontal lines indicate the median response. The gray area (box)
indicates responses between the 25th and 75th percentiles. The whiskers of the plots
(thin horizontal lines) indicate the 2nd and 98th percentile responses. The circles
represent outlier responses of individual participants who fall outside of the 2nd and
98th percentiles. It is interesting to note that only a small number of atheists and
agnostics evaluate the cosmological argument as very strong.

to allow for the following inference: the universe is only finitely old and
started out as a singularity (a very densely packed state). According to this
picture, it does not make sense to talk about time and space prior to 13.75
billion years ago, since time and space began when the universe started to
inflate. This corresponds to the second premise of the Kalām cosmological
argument.

Nevertheless, other interpretations could be given. For example, James
Brian Pitts (2008) argues that, in the case of a singularity, every moment of
the big bang model ($t > 0$) is preceded by earlier moments so that there is
no moment at which the universe began to exist ($t = 0$). This indicates that
scientific evidence itself is not decisive for the cosmological argument but
rather the way in which it is interpreted, and these interpretations depend

on the prior probability one accords to the existence of God, as we saw in the previous chapter.

Some nontheist critics do not accept cosmological arguments because they accord an initially low probability to theism. Materialists may find the existence of the universe as a necessary fact appealing, since their worldview favors the metaphysical primacy of the material world. To them, indeed, there is no reason why the universe, as Hume (1779, 164–165) put it, could not be the necessary self-existent being. Theists, on the other hand, accord a high prior probability to the existence of God and find God an epistemically satisfying explanation. To quote Swinburne (2004, 147), "The choice is between the universe as stopping point and God as stopping point. In the latter case, God's existence and intention over all the universe's history will provide not merely a full explanation but a complete and ultimate explanation of the existence of the universe."

Experimental evidence from cognitive psychology reveals an intricate interplay between explanation and evidence: people treat information as evidence if they can incorporate it in a causal framework (Koslowski et al. 2008). They are more likely to treat background information as relevant to an explanation for an event when the explanation can incorporate both the event and their background information into a single causal account, which makes the overarching explanation increasingly convincing. Merely saying that God created the universe is not as epistemically satisfying as providing a coherent, explanatory account of why we should believe that he did. The cosmological argument can make theism more plausible by its ability to incorporate background information (e.g., the big bang theory) and universal human intuitions about causality and agency into a convincing overarching explanatory account (God as the best explanation for the beginning of the universe).

We have now seen how the acceptance of premises in the cosmological argument and the overall evaluation of the argument's strength critically depend on one's religious outlook. What does this mean for the rationality of religious belief? According to internalist theories of justification, one's belief can be more justified than it previously was if one has proper reasons. This does not mean that beliefs without reasons are unjustified but that additional reasons can further strengthen their justification. Cosmological arguments, like other natural theological arguments, can provide those who already believe based on faith or perhaps on weak evidence with additional reasons for belief. In this way, the cosmological argument can contribute to justified theistic belief in an internalist sense, especially because it starts from premises that humans find intuitively plausible.

The cosmological argument has a positive epistemic function. By its ability to unify scientific knowledge, intuitions about causality and agency, and theism, the cosmological argument contributes to a more coherent worldview in theists, for whom the existence of the universe is a fact in need of explanation. As Craig (2001, 379) puts it, "Even if the uncaused origin of the universe were natural relative to the Standard Big Bang model, that would not imply that the origin of the universe does not cry out for explanation." Kant ([1781] 2005, A624/B652) argued that natural theology is intimately linked with the structure of the human mind: the intuitions that underlie them are subject to human cognitive limitations, but they help to structure our worldview. As discussed in the previous chapter, the persuasive force of the argument from design might be a consequence of its concurrence with human cognitive predispositions, in particular, a universal tendency to see purpose and design in nature. From an internalist perspective, it is therefore possible to treat the intuitions that underlie the cosmological argument as starting points in explanatory strategies that make theism a more coherent position. These intuitions critically depend on intuitive ontologies that are a stable part of human cognition, but they can help us to reflectively evaluate and endorse religious beliefs, thus leading to more justified beliefs.

A worry about this internalist route of justification is what Thomas Kelly (2008) has termed *belief polarization*. This is the tendency by which initially held beliefs become strengthened as one encounters more evidence, even if some of this evidence is not at all in support of the initially held hypothesis (and gets ignored). A large body of research in the psychology of reasoning (see Mercier 2010 for an overview) suggests that this attitude is pervasive: people exhibit confirmation bias, a tendency to selectively attend to and evaluate more positively evidence that is in line with their prior beliefs. They also exhibit disconfirmation bias; that is, they are more likely to dismiss evidence that disconfirms their own beliefs or that confirms beliefs that are incompatible with their own convictions. Unlike some other biases, confirmation bias does not attenuate as a result of schooling; highly educated people are as susceptible to it as those without academic training. An atheist will be more ready to endorse arguments for atheism, whereas a theist will be more inclined to subscribe to arguments for theism. This worry, however, does not necessarily detract from the overall rationality of selectively attending to arguments and their evidential status. It seems sensible for organisms with limited time and resources not to radically overturn their beliefs as a result of each piece of information they encounter, and this may be the reason why this bias is so robust. Once one adopts a position, reflective attention for arguments, even if selective, can provide a more reasoned and, thus, according to internalism, a more justified overall worldview.

Summary

The cosmological argument has enjoyed and still enjoys substantial popularity in various traditions of natural theology. This chapter investigated its cognitive basis. Its enduring appeal is due at least in part to its concurrence with universal human capacities that promote inference to causality and agency. These intuitions form a stable part of human cognition.

Whether or not one can be justified in endorsing the use of such intuitions to argue for the existence of God from the existence of the universe is a different matter. While appeal to intuitions to justify beliefs is a frequently employed strategy in philosophy, such intuitions are vulnerable to evolutionary debunking arguments. However, because causal intuitions in cosmological arguments are similar to those appealed to in scientific reasoning and commonsense reasoning, such evolutionary debunking arguments risk causing collateral damage; that is, they may not only undermine the cosmological argument, but also scientific practice and commonsense reasoning. If these intuitions are so compelling, why do they not convince everyone? As with the design argument, one's evaluation of the cosmological argument depends to an important extent on prior beliefs about the plausibility of theism.

6 The Moral Argument in the Light of Evolutionary Ethics

The idea that theism and morality are inextricably linked has been proposed in theology, philosophy, anthropology, and more recently CSR. Moral arguments for the existence of God take as a starting point this assumed connection between theism and morality. They hold that if objective binding moral norms exist, then God exists. There are several versions of the moral argument, of which we will consider two: the *argument from moral awareness*, which infers the existence of God from human moral intuitions, and the *argument from moral objectivism*, which infers the existence of God from the existence of objective moral norms.

Although the moral argument has not been as popular as some of the other a posteriori arguments for the existence of God, such as the design and cosmological arguments, it has been proposed by, among others, Duns Scotus, William of Ockham, Martin Luther, John Calvin, Immanuel Kant, and more recently C. S. Lewis, George Mavrodes, and Linda Zagzebski. Interestingly, some atheists, such as Friedrich Nietzsche, also accept the link between theism and morality. They argue that objective moral facts are of the same ontological status as God. If it turns out that God does not exist, moral facts do not exist either. Differently put, the same arguments that lead one to deny God's existence should also lead one to reject moral facts.

In this chapter, we will examine the cognitive appeal of premises that underlie the moral argument as a way to assess its cogency. Where do human moral beliefs and intuitions originate? Why do we spontaneously intuit that moral norms are objective and not just dependent upon personal or culturally contingent preferences? Given that such arguments privilege theism as the best or only explanation of altruism, we are interested in the force of moral arguments in the light of evolutionary ethics.

Evolutionary ethics is the scientific study of the evolutionary origins of altruism in humans and other animals. It aims to offer an explanation for why altruistic behavior exists in the natural world and for moral intuitions and emotions. We will first discuss the argument from moral awareness and consider it in the light of evolutionary ethics. Next, we will focus on the argument from moral objectivism. We will examine the cognitive origins of intuitions that underlie this argument and ask whether metaphysical naturalism is compatible with objective moral norms. Finally, we consider whether theism and evolutionary ethics are compatible, and in particular, how a theist might incorporate evolutionary ethics.

The Argument from Moral Awareness

Across cultures, humans have spontaneous evaluative dispositions toward the actions or character of other people. Such attitudes include feelings of liking or disliking a particular person, and evaluating actions as praiseworthy or blameworthy. Moral psychologists have demonstrated that these evaluations occur rapidly and unreflectively. For instance, many Western participants think it is morally wrong for two adult siblings to have a single instance of protected, consensual sex (Haidt 2001). However, when asked why they feel such actions are wrong, many people cannot articulate reasons or arguments. In spite of this inability to justify their moral intuitions, people feel strongly about them: they are not just judgments of taste or matters of individual preference. For centuries, the mind-independent norms that this intuitive moral sense is supposed to track were collectively referred to as *natural law* (*lex naturalis*). The origins and properties of the natural law have been the object of intense philosophical and theological debate. Traditionally, theologians located its origin in God. For instance, Aquinas argued, "The natural law is promulgated by the very fact that God instilled it into man's mind so as to be known by him naturally" (Aquinas [13th c.] 2008, 35).

Humans have what Richard Joyce (2006, chapter 4) has termed a *moral sense*, a set of innate cognitive capacities that prompts them to make judgments in the moral domain. The argument from moral awareness holds that this moral sense tracks a mind-independent natural law and that its properties are best explained by the existence of God. Swinburne formulates the argument as follows:

If God is to give us significant choices, he will ensure that we will develop this kind of moral awareness. But if there is no God, how likely is it that embodied creatures with a mental life will progress to this stage? If genetic mutations produce creatures naturally inclined to behave altruistically towards others of their community ... then there may well be a good Darwinian explanation for their survival. ... Such creatures may help each other spontaneously and naturally, as many groups of animals do. But having the understanding of these actions as morally good (even when we do not desire to do them) is something beyond mere altruistic behaviour. And there seems no particular reason why any mechanisms of mind–body interaction that cause creatures to have beliefs should produce moral beliefs. (Swinburne 2004, 216)

John Hare (2004) also holds that natural selection cannot explain the moral sense. Following Duns Scotus, he argues that humans have an affection for justice (a sense of moral obligation) and an affection for advantage (self-regarding tendencies). Evolutionary theory predicts that we rank the affection for advantage higher than the affection for justice. Why then should we feel moral obligations if evolution has produced the wrong ranking? Hare suggests that divine grace explains how we can get our rankings right, that is, place our affection for justice first.

Arguments of this kind, which rely on the (alleged) failure of scientific psychological theories to explain the moral sense, predate the emergence of evolutionary ethics. For instance, C. S. Lewis ([1952] 2002, 10–11) thought that "instincts" (the ethological explanation for behaviors prevalent in his time) do not explain the intuitive awareness of morality that humans have because instincts tell us nothing about their relative importance or ranking in specific situations. The argument from moral awareness is an argument to the best explanation, roughly of the following form:

1. To explain the moral sense, we can invoke either theistic or naturalistic explanations.

2. Naturalistic explanations cannot account for the moral sense.

3. Theism can account for the moral sense.

4. Therefore, the best explanation for the moral sense is theism.

One immediate worry is that the argument from moral awareness, by invoking premise 2, is a God of the gaps argument, which is often regarded as weak or even fallacious. Robert Larmer (2002) contends that not all God of the gaps arguments are weak. They can be fallacious if they are of the form "We don't know the natural origin of x, so God caused x," where x stands for properties like complexity, consciousness, or moral awareness.

In this form, the moral argument becomes an argument from ignorance. By contrast, God of the gaps arguments can be stronger if they rely on an inference to the best explanation. According to this form of the argument, it is reasonable to hold that God—if he exists and is supremely good—would create beings with a moral sense. This explanation is not available to metaphysical naturalists. However, Swinburne, Hare, and Lewis want to argue for something stronger: they propose that natural explanations are *in principle* unable to account for moral intuitions and beliefs. Each of these arguments relies on a specific scientific framework (evolutionary theory in the case of Swinburne and Hare, ethology in the case of Lewis), so it is hard to see how this assumption can be generalized. Perhaps it is more prudent to consider a weaker version of the argument from moral awareness, where theism is regarded as the best explanation for human moral intuitions, without, in principle, excluding naturalistic explanations. In the next section, we will consider whether this version of the moral argument is cogent by exploring naturalistic explanations for human morality.

The Evolution of Human Morality

Biological Altruism
Many evolutionary ethicists (e.g., James 2011) take altruistic behavior in nonhuman animals as a starting point. We will here use the term *biological altruism* to denote behaviors or dispositions that increase the fitness of other individuals and that are costly to the organism exhibiting this behavior. In this technical sense, many forms of biological altruism do not require altruistic sentiments such as compassion and empathy. For example, cleaner wrasses (*Labridae*) are a family of small fish that clean parasites and dead tissue from the skin of larger fish. This behavior is altruistic in the sense that evolutionary biologists use this term; cleaner wrasses probably do not feel any sympathy or concern for their hosts' wellbeing.

One of the enduring puzzles in evolutionary biology is why biological altruism persists, given that it seems to result in fitness costs to organisms that exhibit such behavior. Evolution is based on competition between individual organisms, which leads one to expect that animals would behave purely selfishly. Yet cooperation is widespread. For example, Belding's ground squirrels (*Spermophilus beldingi*) emit alarm calls when they see a predator. This behavior helps group members but poses an increased risk to the calling individual (Sherman 1977). Since the 1960s and 1970s, most typical instances of biological altruism have been explained by the following mechanisms: kin selection (Hamilton 1963), direct reciprocity (Axelrod and Hamilton 1981), indirect reciprocity (Nowak and Sigmund 2005), and group selection (Bowles and Gintis 2011).

Kin selection conceptualizes fitness not from the point of view of the organism but from that of the gene. An organism that helps its kin in an altruistic way decreases its individual fitness but increases its inclusive fitness, that is, the probability that copies of its genes in closely related individuals are spread. Belding's ground squirrels mainly interact with direct kin. Hence, individuals in the immediate vicinity of the caller tend to be closely related. The latter increases its risk of being eaten but also augments the probability that its siblings or offspring (typically sharing 50 percent of its genes) will survive. If any of these individuals also carry the calling gene, this has a higher chance of survival and can spread throughout the population.[1]

Direct reciprocity follows the tit-for-tat strategy ("I help you, you help me"): animals that are not close relatives can benefit from being altruistic if the other returns the favor. For example, pied flycatchers are more likely to help a conspecific mob a predator if that individual helped them before (Krams et al. 2008).

Indirect reciprocity relies on the principle "Since you helped someone else, I'll help you." It works primarily by reputation, where altruistic acts increase an agent's reputation and selfish acts decrease it. Agents with a reputation for being generous are more likely to receive favors from others. Given that this requires keeping track of the behavior of many agents, this form of altruism relies on gossip (who did what for whom?), and therefore may require language. Martin Nowak (2006) speculates that indirect reciprocity was an important driving force in the evolution of human morality and social norms.

Group selection, while still controversial, is making a comeback. For group selection to work, groups must be distinct from each other and form cohesive wholes. It requires that the fitness benefits of altruistic groups over selfish groups outweigh the fitness benefits of selfish individuals over altruistic individuals within mixed groups. Once stable groups have been formed, altruistic behavior can be favored within such groups, and their members will have higher reproductive success compared to members of less altruistic groups. Human cultures, with their differences in language, clothing, norms, cuisine, and other distinguishing features, may be good candidates for group selection: they exhibit high between-group variation and considerable within-culture homogeneity (Henrich 2004; De Smedt and De Cruz 2012).

Psychological Altruism

As we have seen, several forms of biological altruism observed in humans and other animals can be explained without attributing any feelings of empathy or concern to them. Yet, humans (and maybe other animals too) frequently have emotions that are other-directed, such as empathy and concern. We will refer to such emotions collectively as *psychological altruism*. Evolutionary ethicists propose that psychological altruism evolved to facilitate biological altruism. Ultimately, what drives natural selection are behaviors, real-world interactions of an organism with its environment. When the fitness outcomes of a behavior depend on context, mental capacities, such as emotions and reasoning skills, are needed to help an organism act adaptively. Animals that live in groups with complex social structures and interactions cannot rely solely on reflex-like responses but require flexible and context-sensitive behavior that keeps track of how others behaved in the past, which explains the evolution of psychological altruism. For example, a feeling of resentment is a powerful negative emotion directed toward someone who has harmed (or is perceived to have harmed) the person who feels it. This emotion can guard one against being taken advantage of.

Frans de Waal (2009) contends that many nonhuman animals, especially monkeys and apes, have altruistic sentiments such as empathy, as can be inferred from their complex prosocial behavior, which includes reconciliation, spontaneous helping, and consolation. However, because altruistic sentiments are phenomenological states, it is difficult to establish psychological altruism conclusively in nonhuman animals. To take a recent example, Bartal, Decety, and Mason (2011) showed that rats will spontaneously free a conspecific that is trapped in a tube as it struggles and emits shrieks of distress. The authors interpret this as evidence for empathy in rats, but the behavior is consistent with alternative explanations, including that the subject wants to end its own distress caused by hearing the continuous shrieks of its fellow, or even that the helper wants to get into the tube. (Several rats in fact entered the tube after freeing their cage mate.) While it remains unclear to what extent psychological altruism exists in nonhuman animals, it seems reasonable to assume that it evolved in humans through natural selection as a way to secure (more complex forms of) biological altruism, such as indirect reciprocity.

The Moral Sense

We will now turn to the moral sense, the human capacity to form moral evaluative attitudes, which is the subject matter of the argument from moral awareness. Humans frequently make moral judgments. Joyce (2006) proposes that humans have an innate moral sense, which underlies our ability to make such judgments. This capacity does not require that moral-ity with a particular content is innate but, rather, that we spontaneously regard and appreciate some actions or persons as morally good or bad. There is at present a lively discussion on when and how the moral sense arose. Some authors, such as de Waal (2009), discern a broad continuity between human morality and psychological altruism in nonhuman animals. At the other extreme, Francisco Ayala (2010) hypothesizes that the capacity for morality arose late in human evolution as a byproduct of recently evolved human reasoning capacities. Most evolutionary ethicists (e.g., Silk and House 2011) occupy an intermediate position: while they acknowledge psychological altruism in nonhuman animals, the moral sense has unique features. In particular, it has a prescriptive, nonsubjective character: we do not just abide by moral norms because we feel like it, but because we feel we ought to do so.

Since animals cannot talk about their internal mental states, evidence for a moral sense that is not preference-dependent can only be obtained indirectly, for instance, by studying behavior in the wild or under labo-ratory conditions. Based on such observations, there is an emerging con-sensus that animal moral motivations are more preference-dependent than those of humans. For example, chimpanzees as well as other primates rec-ognize when they are unfairly treated (e.g., when they receive a smaller reward than a conspecific) and show displeasure when this happens. How-ever, they are not averse to exchanges where they benefit more than others. Indeed, when given the opportunity to either deliver a food reward only to themselves, or a reward to themselves and to a familiar (unrelated) indi-vidual, they are equally likely to choose the selfish as the altruistic option, even though the altruistic option does not cost them anything (Silk et al.

2005). By contrast, cross-cultural experiments indicate that humans are ill-disposed toward inequality even if they do not directly suffer from it: they are willing to punish someone, even if punishing is costly, if they think that that person has treated someone else unfairly (Henrich et al. 2006).

When does sensitivity to moral properties arise? Empirical studies suggest that infants can recognize and evaluate morally relevant behaviors such as helping and hurting. In one experiment, six-month-olds were shown two events (Hamlin, Wynn, and Bloom 2007). In both cases, a red circle attempts to climb a steep hill. In one of the events, another geometric figure "helps" the circle by pushing it uphill; in the other event, another geometric figure "hinders" the circle by pushing it downhill. Infants were subsequently presented with the helper and hinderer shapes and invited to pick one. The vast majority of infants preferred the shape they saw previously helping over the figure they saw hindering. In a more recent follow-up study (Hamlin et al. 2011), toddlers preferred to take away the cookie of a puppet they previously saw interacting in a negative way over the cookie from an agent they previously saw helping someone out. They also avoided taking the last cookie of the puppet they witnessed being harmed. The authors interpret these findings in a rich way, as meaning that these children have a moral sense like adults, but it is not clear what moral intuitions and beliefs should be invoked to explain their reactions.

Neuroscientific research indicates that our ability to form moral evaluative attitudes strongly depends on emotion-related brain areas involved with reward and other-directed feelings, such as the thalamus, anterior insula, and ventromedial prefrontal cortex, as well as reasoning-related areas, such as the dorsolateral prefrontal cortex and anterior cingulate cortex (see Moll, De Oliveira-Souza, and Zahn 2008 for review). For example, when people reject an offer they perceive as unfair, the decision is preceded by increased activity in the anterior insula, a brain area strongly connected to emotional response. People with damage in the ventromedial prefrontal cortex can still verbally state the moral norms of their society but have difficulties making practical decisions that apply this knowledge in particular situations, leading them to make suboptimal decisions.

Evolutionary ethicists such as Michael Ruse (1995) hypothesize that the moral sense evolved as a way to help maintain biological altruism. The problem with this explanation is that psychological altruism is already supposed to do this (Lahti 2003). If other-regarding feelings already motivate us to cooperate, why would we need an additional moral sense?

Some authors do not think the moral sense is innate. Rather, they believe our propensity to make moral judgments is a cultural offshoot of psychological altruism: there is no separate domain of evolved moral cognition; rather, there is a cultural domain of moral cognition that builds on psychological altruism. Because of their attempt to reduce morality to emotions, such theories are termed *sentimentalist theories of morality* (e.g., Prinz 2006). Although moral norms are cultural constructs, they do not change in a purely arbitrary fashion; their transmission is guided and constrained by our emotions (Nichols 2004). Norms that resonate with our emotions are more likely to be culturally transmitted; those that leave us emotionally indifferent are more likely to disappear. Take the cultural evolution of harm norms. Harm norms limit or prohibit causing pain and suffering in others. Although almost all documented cultures impose some limit on harm, there is considerable variation in harm norms across cultures. In the course of European history, harm norms have become stricter, especially since the Enlightenment: flogging children is now seen as morally impermissible but used to be regarded as a proper part of child rearing; wanton animal cruelty (e.g., throwing stones at cats) was once innocent child's play but is now unacceptable. Shaun Nichols explains this shift in moral norms by appealing to our overwhelmingly negative emotional responses to suffering in others: watching others in pain is painful. Neuroimaging studies by Jackson, Meltzoff, and Decety (2005) indicate that the visual perception of painful situations (e.g., fingers stuck between a door) activates a large network of brain areas involved in processing one's own pain—there is a neural overlap between the perception of one's own pain and that of others, which may explain why it is unpleasant to see another person being physically harmed. The moral norm of not harming others may have arisen through a culturally contingent process (the moral values of the Enlightenment), but once it arose, it was likely to be transmitted because it is in accordance with our evolved emotional responses. A problem with sentimentalist theories is that they do not explain why moral norms develop reliably across cultures. There is cultural variability, but the *existence* of moral norms is a cultural constant. If moral norms are merely cultural byproducts of emotions, why do we not find cultures without moral norms? Also, sentimentalism does not adequately explain why people perceive moral norms as binding and obligatory. Our emotions caused by perceiving pain and suffering in others may lead us to be attracted to cultural norms that limit harm, but this fails to explain why we think that we ought to abide by these rules.

The hypothesis that the moral sense is a biological adaptation thus seems plausible. In humans, cooperation is more extensive than in other primates, so it seems probable that humans need additional mechanisms to secure this high level of cooperation, for instance, in their unprecedented levels of cooperative breeding and cooperative hunting (De Smedt, De Cruz, and Braeckman 2009; De Smedt and De Cruz 2014). These require psychological mechanisms that secure cooperation even if it goes against the organism's immediate self-interests. By curbing self-interested behaviors, a moral sense can secure long-term adaptive benefits. Once hominids developed a brain size that was markedly larger than that of chimpanzees, which occurred about two million years ago, they were forced to wean their infants earlier to meet the nutritional demands of the rapidly growing brain (Kennedy 2005). This led to shorter interbirth intervals compared to other apes. At the same time, human children have a long period of dependency, needing close supervision and care until they are six or seven. This places heavy demands on human mothers, who care for several dependent children at the same time. In many small-scale societies, women solve this problem by developing extensive networks of alloparents. In some cases, as with the Aka pygmies, childcare networks comprise more than ten individuals including several nonkin members (Ivey 2000). To do this successfully, moral norms about taking care of the children of others have to be obeyed, and temporary self-interests have to be curbed. Benoît Dubreuil (2010) conjectures that humans as early as *Homo heidelbergensis* (ca. 700,000 years ago), the common ancestor of *Homo sapiens* and *Homo neanderthalensis*, evolved the capacity to stick to moral norms that promote altruism in the face of conflicting selfish motivations. *Homo heidelbergensis* engaged in high-stake cooperative activities such as large-game hunting, where defection was costly for all parties. Groups of hominids that were able to stick to moral norms had higher fitness than groups that did not follow such norms, which accords with the propensity of current-day humans to follow norms even in the face of competing self-regarding motivations.

Combined with this evolved sense of moral obligation, cultural factors may have played a role in specifying the content of moral norms in a process of gene-culture coevolution. Coevolutionary scenarios (e.g., Allison 1992) indicate that moral norms such as "love your neighbor" and "respect your elders" have a high chance of being maintained because such norms increase the average fitness of members of a group. Caring for one's neighbor increases cooperation; having respect for one's elders helps the transmission of valuable knowledge from experienced members of the community, especially in environments that are relatively stable. People need not be aware of the fitness benefits of such norms. Under particular cultural and socioeconomic circumstances, cultural evolution can maintain highly specific systems of moral rules, such as the norms of Greek hoplite warfare, which involved avoiding shame (by not breaking ranks) and dedicating the spoils of war to the gods (Runciman 1999). Such cultural models tell little about the individual motivation of people to obey particular moral norms, but combined with evolutionary models that explain the human propensity to follow moral norms, they may explain why humans can form moral evaluative attitudes and why they believe that moral norms trump individual interests.

Currently, naturalistic explanations for morality seem to be in a sketchy state, whereas theistic accounts (e.g., divine command theories) are well developed.[2] Nevertheless, they indicate that it is not in principle impossible to explain in a naturalistic way the capacity to evaluate and follow moral norms. As we saw, Swinburne, Hare, and Lewis state that naturalism will never be able to explain moral awareness (premise 2 of the argument of moral awareness). Although as of yet, there is no complete naturalistic explanation, some naturalistic accounts encroach upon the terrain that theists have carved out for themselves. Premise 2 seems too strong given the recent developments in evolutionary ethics.

The Argument from Moral Objectivism

Many moral arguments assume that the existence of objective moral norms favors a theistic worldview. Unfortunately, there is no agreement on what exactly "objective" means. Philosophers usually mean by this that it is non-subjective, that is, not dependent on one's individual desires, emotions, or preferences. Not everyone goes as far as to equate objective to absolute, that is, not dependent in any way on human reasoning, social contracts, or institutional facts. In this latter, more restrictive sense, objective means either that morality depends on facts in nonhuman nature (e.g., evolution) or on nonnatural properties (e.g., abstract moral facts that exist independent from the natural world). Using objectivism in this nonnaturalistic sense, Robert Adams (1987) provides an argument from moral objectivism along the following lines: moral facts are both objective and nonnatural, properties that are best explained by theism. Therefore, we have good reasons to think that theism is true. Some versions of the argument from moral objectivism are deductive, as in Paul Copan's (2008) formulation:

5. If objective binding moral facts exist, then God exists.

6. Objective binding moral facts exist.

7. Therefore, God exists.

As we can see, the argument from moral objectivism assumes that there are objective, mind-independent moral facts, and it strongly favors a nonnatural interpretation of them.

Moral antirealism argues that moral statements are not mind-independently true or false. As Nietzsche ([1889] 2005, 182–183) succinctly put it, "*There are absolutely no moral facts*. What moral and religious judgments have in common is the belief in things that are not real. Morality is just an interpretation of certain phenomena or (more accurately) a *misin*terpretation" (emphasis in original). Antirealism does not deny that moral intuitions play a role in motivating people's actions, nor does it deny that we *perceive* some ethical norms to have an objective force. It is even consistent with the view that moral norms can be conducive to human survival, well-being, and flourishing. What it does deny is that moral claims can be true in some categorical, noninstrumental sense. It makes no sense for a moral antirealist to say that providing aid to starving children is something that is good in itself, and not just something that helps those children survive (Sommers and Rosenberg 2003). Nihilism, a specific form of moral antirealism, holds that statements such as "slavery is wrong" are either false or meaningless; we may think the statement has an objective moral value, but we are mistaken (Mackie 1977). Proponents of the argument from moral objectivism (e.g., Copan 2008) freely acknowledge that their argument does not work if we do not accept moral realism.

The argument from moral objectivism relies on two assumptions: objective moral norms exist (in the strong, nonnaturalistic sense), and there is a link between objective moral norms and the existence of God. In the next sections, we will argue that the plausibility of these premises has a cognitive basis: humans intuitively think that moral norms are objective and that there is a link between these and theism.

Is Moral Realism Intuitive?

Philosophers have widely assumed that moral realism is the commonsense view. Interestingly, they seem to agree on this regardless of whether they subscribe to this position. For proponents of the moral argument (e.g., Rogers 2005), commonsense moral objectivism lends support to its prima facie plausibility. By contrast, moral antirealists (e.g., Ruse 1995) think that intuitive moral realism is an illusion. Despite these differing interpretations, realists and antirealists seem in striking agreement about the commonsense nature of moral realism. Does this view stand up to empirical scrutiny? Ruse and Wilson (1986, 179) hypothesize that humans are intuitive moral realists and that this metaethical intuition has an evolved function: "Human beings function better if they are deceived by their genes into thinking that there is a disinterested objective morality binding upon them, which all should obey." If everyone thought that morality was subjective, that it was merely a matter of taste or convention, our social systems would collapse. Intuitive moral realism is thus a key component in human altruistic behavior, held together by moral beliefs, which in turn are cemented by an intuitive moral realism. As Ruse (2010, 310) wrote later on, "Substantive morality stays in place as an effective illusion because we think that it is no illusion but the real thing."

When Ruse first formulated this hypothesis, it was by no means clear that humans were intuitive moral realists. Also, it was unknown to what extent an intuitive moral realism helps us to act more morally. In the meantime, there is some empirical work on this. Preschoolers already make a distinction between moral and conventional norms (Nucci 2001). They regard moral violations as less contingent upon authority and more generalizably wrong than violations of social conventions. For instance, they think that it is okay to chew gum in class if the teacher allows it, but that it is still wrong to hurt other children, even if he condones it. Nichols and Folds-Bennett (2003) found that preschoolers see moral properties as independent from individual preferences, in contrast to taste properties, which are preference-dependent. Children were also more likely to intuit that moral properties generalize across cultures (e.g., hitting and pulling hair is also not allowed in other countries) than they were to think that taste properties generalize across cultures. Strikingly, Goodwin and Darley (2008) observed similar strong realist tendencies in Western students: participants thought that moral statements were almost as objective as scientific facts and more objective than social conventions or expressions of taste. Young and Durwin (2013) used an implicit test to gauge the motivating force of moral realism. They set up an experimenter as a street-canvasser for a charitable organization. Passersby were primed with either moral realism or antirealism. Those primed with moral realism were twice as likely to donate money compared to those from the antirealism or a neutral control condition. While these studies support Ruse's idea that moral realism is a folk metaethical position and that it motivates us to behave morally, they do not prove that moral realism is an evolved cognitive illusion. Indeed, these findings are compatible with the view that folk moral realism is adaptive because it tracks a real-world property, the reality of moral facts. To establish that moral realism is an illusion, one would have to show that moral realism is false. We will return to this question later in this chapter with a discussion of evolutionary debunking arguments against moral realism.

CSR and the Link between Theism and Morality

Cross-culturally, religious beliefs and practices seem to help members of religious communities to overcome problems inherent in human cooperation, such as the risk of free riders. For example, a literature study on the viability of nineteenth-century American communes found that religious communes had on average a lifespan that was four times longer than that of secular (ideology-based) groups (Sosis and Bressler 2003). As communes tend to fall apart because of increasing levels of free riding, this suggests that religious settlements were more effective in maintaining high levels of cooperation. In what follows, we will review evidence and theories from CSR that examine the link between theism and morality.

As we saw in chapter 3, children start out overattributing omniscience and only later restrict this to God. However, our concept of supernatural omniscience is often tied up with moral sanctions. Humans are not so much interested in whether God knows what is inside some box but, rather, whether he knows who behaves morally well and who does not. This type of information about socially relevant actions in the moral domain is termed *socially strategic information*. Boyer (2002) hypothesizes that belief in supernatural agents who have access to socially strategic information is culturally widespread because this information matters a great deal to us. Our reputation determines in how far we can enjoy the benefits of indirect reciprocity. Socially strategic information can be positive (e.g., someone has displayed an exceptional level of generosity), but is often negative (e.g., someone has broken her promise or cheated). We use this information to gauge the reputation of others and to decide whether or not to cooperate with them, an important condition for indirect reciprocity. Agents who have a privileged knowledge of actions that affect reputation are socially relevant, which makes them memorable and thus likely to be culturally transmitted. Purzycki et al. (2012) asked participants whether God and other agents knew about nonstrategic information (e.g., if God knows the number of moons around Mars), or socially strategic information (e.g., if God knows that Ann gives to the homeless). They found that responses for socially strategic information were faster than for other information. The fastest response times were for negative socially strategic information (e.g., if God knows that John cheats on his taxes). It turns out that we especially care about other agents' knowledge about morally relevant actions if such beings punish or reward: Purzycki et al. (2012) found that participants took less time deciding whether an agent knew about some socially strategic information if the agent was able to punish. For example, people's response patterns were similar for God as for a nonsupernatural governmental agency with cameras everywhere that could punish transgressors. If the agent was not interested in morally relevant information or was unable to punish, there was no difference in reaction times. For example, there was no difference in the responses about knowledge (socially strategic or nonstrategic) possessed by aliens who observed all human actions but did not interfere with human affairs.

What kind of supernatural being has access to socially strategic information and is willing and able to punish those who transgress moral norms? Anthropological studies paint a rich picture of a variety of watchful agents, including ancestors, spirits, and gods. People who think they are being watched behave more prosocially than those who believe they are anonymous. For example, students are more likely to put money in a donation box at a self-serve coffee station (to pay for what they take) when it is

highlighted by two staring eyes (Bateson, Nettle, and Roberts 2006). Likewise, Bering et al. (2005) found that participants were less likely to cheat at
a computer game where they could win money if, prior to the experiment,
they had been informed that the ghost of a deceased graduate student had
been observed in the room the experiment took place in.

Given that the belief that we are being watched reduces antisocial behavior and promotes altruism, belief in moralizing supernatural beings would
have proven particularly effective at stabilizing large groups. In the course
of human history population size increased and interactions between people became more and more between strangers and less between family and
friends. In a large group, where members do not know each other on a
personal basis, reputation monitoring (such as gossip) is no longer effective, threatening its stability and increasing opportunities for free-riding. If
its members believe in watchful and punishing supernatural beings, group
stability can be better maintained. According to Shariff, Norenzayan, and
Henrich (2010), *high gods* are very effective in enforcing moral behavior
within groups. High gods are super-knowing, powerful, morally concerned
supernatural beings. Belief in high gods is indeed restricted to large-scale
societies, probably because small-scale cultures can maintain cooperation
by direct and indirect reciprocity (Roes and Raymond 2003). Their historical emergence is strongly correlated with the development of sedentary, agricultural, large groups, where members do not know everyone on
a personal basis (Sanderson and Roberts 2008). The intuitive connection
between theism and morality that forms the basis of the moral argument
may thus find its origin in our cultural history: Western theists (and nontheists) have been raised in large-scale societies that historically (or still)
hold belief in a high god.

Still, this does not entirely explain the connection between theism
and morality. In one of the few cross-cultural studies that have explicitly
addressed the moral concerns of non-high gods, Ben Purzycki (2013) examined the knowledge and concerns that Tyvans (Buddhists and shamanists
living in western Mongolia) attribute to their spirit-masters, local supernatural beings that inhabit specific places. When explicitly asked about this,
Tyvans hold that the spirit-masters are not concerned with socially strategic
information and that these beings do not punish transgressors. Nevertheless, as is apparent from their answers, implicitly, believers think that the
spirit-masters are more knowledgeable about and more concerned with this

type of information (e.g., whether or not someone lied) than with other facts (e.g., what is growing in one's yard). The Tyvans show a similar moralizing bias as Christian participants, who believe God knows more about morally relevant facts than about other facts. This is an intriguing finding, but it is not incompatible with a cultural origin of the link between theism and morality. Humans may find agents who are concerned with morality more interesting and may intuitively believe that supernatural beings are concerned with moral transgressions, hence the Tyvans' implicit beliefs about the spirit-masters' interest in socially strategic information. However, since the spirit-masters do not punish, they cannot keep people in line and thus cannot become effective promulgators of morality. This is in accordance with the explicit beliefs Tyvans hold about them.

The empirical and theoretical work in CSR reviewed in this section can shed light on premise 5 of the argument from moral objectivism (if objective binding moral facts exist, then God exists). Westerners tend to think about religion and morality as linked because they stem from cultures with a high god. This is also the case for atheists such as Nietzsche who are heirs to the view that belief in God keeps people morally in line. If there is no high god, why stick to moral norms? Nietzsche strongly suspected that there is no motivation or reason for doing so. In the next section, we will look at contemporary debunking arguments against moral realism that explore similar ideas.

Evolutionary Debunking Arguments against Moral Realism

Evolutionary debunking arguments (EDAs) against moral realism have given rise to lively philosophical discussion. They are an intriguing mirror of moral arguments. The latter start out with the premise that there are objective moral norms and, therefore, theism is true. EDAs against moral realism, by contrast, start out with the premise that metaphysical naturalism is true and conclude from this that there are no objective moral norms. In what follows, we will review evolutionary debunking arguments by Richard Joyce (2006) and Sharon Street (2006) and examine their implications for the moral argument.

EDAs against moral realism tend to combine several strategies. A common pathway is to argue that even if mind-independent moral truths existed, our evolved minds would be insensitive to them. Our minds tend to form moral beliefs that are conducive to reproduction and survival in our social milieu, not beliefs that track moral truths. According to Joyce (2006), this is the crucial difference between moral beliefs and, for instance, arithmetical beliefs. Unlike the latter, moral beliefs may be useful even if they are false. Although this argument cannot rule out that mind-independent moral truths exist, it aims to show that such truths are redundant for explaining our beliefs. Therefore, it is more parsimonious to assume that there are no moral truths.

A concrete example offered by Street is the moral intuition that suffering is bad. When we think that suffering is bad, we rely primarily on an evolved aversion to pain in ourselves and in others. Its evolution can be explained without attributing any objective truth-value to the badness of pain. Even if pain were good in some objective moral sense, we would believe it to be bad because such a belief enhances fitness, as it would motivate us to avoid pain. Therefore, our moral intuition that pain is bad is parsimoniously explained by our evolved aversion to pain, and there is no need to invoke a mind-independent moral truth that pain really is bad. Indeed, even if moral truths existed, it would be unlikely that natural selection led us to uncover them. To use Street's (2006, 121) striking metaphor, the chance of that happening is as remote as someone who sets sail hoping that the winds and tides will bring her to Bermuda, to actually arrive safely at her destination. Natural selection will only steer us toward holding moral beliefs that are adaptive for those who hold them.

Not everyone agrees that metaphysical naturalism is incompatible with moral realism. A crucial premise of EDAs is that our moral beliefs track facts that are important from an evolutionary point of view and that they therefore do not track moral truths. One can challenge this assumption by making a link between evolved moral beliefs and mind-independent moral truths. To David Copp (2008), our evolved moral intuitions track moral facts, just like our evolved beliefs about the danger of predators track facts about the world. He proposes a form of social functionalism, according to which the well-functioning of a society depends on its stability, cooperation among its members, and friendly relations with neighboring groups. Groups that have moral norms that promote these features (e.g., do not covet your neighbor's wife) do, on the whole, better than groups that do not have such moral norms. Thus, cultural evolution will favor the emergence of moral intuitions and norms that promote behaviors that benefit human groups. Through a process of gene-culture coevolution, people whose evolved moral intuitions track these institutional moral facts will have higher fitness. Hence, our moral beliefs are not purely subjective but reflect moral truths that are located in institutional facts, a view that harkens back to John Dewey (1898), who speculated that ethics could arise through a form of social selection.

Scott James (2011) proposes a similar view, where moral truths are founded in the evaluative attitudes of others. More specifically, people who developed a sensitivity to whether or not others could reasonably object to their conduct had higher fitness than those who did not. James argues that this moral sense provides us with an ability to successfully interact with others. This sensibility allows for higher levels of cooperation than if we had only access to psychological altruism. For one thing, altruistic sentiments do not help us to adjudicate between conflicting moral demands, whereas the moral sense does. These proposals are in line with the earlier outlined evolutionary hypothesis that the moral sense developed in human evolution as a result of increased cooperative behavior (alloparenting and large-game hunting) through gene-culture coevolution.

Although moral realist positions like these are still germinal, they indicate that those who argue that metaphysical naturalism is incompatible with moral realism are moving too quickly. Theism has at present a better account for moral realism—it is indeed easier to envisage the existence of objective moral norms in a theistic worldview. But it may be premature to compare both positions: theistic explanations of morality have been on offer for more than 2,400 years, whereas naturalists have only seriously started tackling morality since the advent of sociobiology in the 1970s, a few exceptions such as Darwin (1871a) and Dewey (1898) notwithstanding. Nevertheless, it is not easy to envisage how the metaphysical naturalist can argue that these evolved objective moral norms could be *binding*. A descriptive naturalistic claim cannot straightforwardly move toward a prescriptive, normative claim. Naturalists can use other principles to get from is to ought, such as an appeal to the objectiveness of harm. As Walter Sinnott-Armstrong (2009, 101) argues, "What makes it morally wrong to murder, rape, steal, lie, or break promises, for example, is simply that these acts harm other people without any adequate justification." However, while the naturalist can argue that such harm is bad, it is not clear why it would be *wrong*. Wrongness cannot be extracted from, or justified by, factual claims about human moral norms.[3]

Are Theism and Evolutionary Ethics Compatible?

Having considered the viability of the moral argument in the light of evolutionary ethics, we will now examine whether a theist can coherently believe that moral intuitions are the result of a naturalistic evolutionary process. One way to combine theism and evolutionary ethics is to assume that God constructed morality through evolution. Such a picture may be subject to an evolutionary version of the Euthyphro dilemma. The classical Euthyphro dilemma (originally formulated by Plato) asks whether something is good because God wills it, or whether God wills it because it is good.

Evolutionary Euthyphro dilemma

Either: the moral good is good because God has instilled the appropriate moral sentiments through natural selection.

Or: God recognizes and commands as morally good properties that have evolved through natural selection.

The first horn of the dilemma presents the evolutionary theist with the problem of arbitrariness. If human evolution had gone in a different direction, our moral intuitions would probably have been different. Darwin (1871a, 73) hypothesized that if humans had evolved eusociality, "our unmarried females would, like the worker-bees, think it a sacred duty to kill their brothers, and mothers would strive to kill their fertile daughters; and no one would think of interfering." If we were eusocial, we would place much less value on individual lives and be more concerned with the survival of the hive or nest. In fact, eusociality is quite common in nature, as it has evolved three to eleven times independently (depending on how it is defined) in such disparate clades as insects, shrimps, and mammals (West and Gardner 2010). To us, it is wrong for women to kill their fertile offspring, but it would have been right had we lived in a eusocial structure, or it would be right for eusocial animals that had evolved a full-fledged moral sense. This may be unpalatable to those who hold that human moral values have a special objective status. Evolutionary theists who are willing to grasp the first horn can argue that, as a matter of fact, humans are not eusocial.

Unlike metaphysical naturalists, they need not accept that evolution is a purely contingent process. Indeed, in the Christian tradition, the creation of humans is part of God's plan. This does not imply that theists should endorse something like a guided theistic evolutionary process, which God, through direct intervention, nudged in such a way that the human mind became equipped with the moral intuitions it has. This guided form of theistic evolution, perhaps most famously proposed by Teilhard de Chardin (1959), has current adherents as well (e.g., Plantinga 2011). Also, theists need not endorse a form of convergent evolution in which intelligent, sentient beings with moral sensibilities are the inevitable result, as suggested by Simon Conway Morris (2003). This is not to say that Conway Morris's view is incorrect, but it is controversial, and theists need not be committed to it. Rather, they can interpret the stochasticity of evolution as part of God's plan. The role of chance in evolution does not necessarily entail a lack of purpose. To take an analogy: lotteries are genuinely stochastic, and in this stochasticity, they serve a global purpose, namely making money for those who organize them—this purpose is achieved regardless of the specific outcome of the lottery, that is, whoever wins the grand prize (Hall 2009). Similarly, the undetermined evolution of beings through natural selection may have as global outcome the evolution of free-willed, morally conscious beings. God may have left the precise details of how this is achieved up to chance in order to make his creation genuinely free (see, e.g., Haught 2000; Miller 2007 for such views).

The second horn of the dilemma presents the problem of sovereignty: if evolution is capable of generating moral properties (as suggested in previous sections), which God then recognizes, why invoke theism at all? Evolutionary theists willing to grasp the second horn may argue that not all evolved moral inclinations are good and, hence, that not all are endorsed by God. For example, humans universally care more for and are more altruistic toward relatives and members of their in-group (xenophobia). Humans have also a natural tendency for differentiation by wealth or social status. This is already a feature of human groups since at least the Upper Paleolithic, as can be seen in lavish burials with rich grave gifts for some individuals as compared to others (e.g., Vanhaeren and d'Errico 2005). A theist could argue that God only commands a subset of the evolved moral intuitions. Our evolved moral intuitions provide the raw material from which cultural moral norms are constructed. Through revelation or the workings of the Holy Spirit, God can still communicate which moral intuitions we should listen to (e.g., do not harm others) and which ones we should perhaps ignore (e.g., nepotism). Given that human societies change over time, we should perhaps expect that the truth-value of moral claims could also change. In societies today, favoring members of the in-group over members of the out-group is arguably less morally defensible than it was in prehistory, when most interactions took place between in-group members, and when dealing with people outside of one's own group was mediated by social mechanisms like reciprocal gift-giving (Whallon 2006). Such mechanisms still play a role in current small-scale societies (Apicella et al. 2012) but are morally dubious in large-scale societies, where daily interactions are not restricted to in-group members or relatives.

Summary

Proponents of the moral argument and some evolutionary ethicists have argued for a close connection between theism and ethics. Defenders of the moral argument contend that if objective moral norms exist, metaphysical naturalism is false (the argument from moral objectivism). The argument from moral awareness holds that the human moral sense is inexplicable within a naturalistic framework.

In this chapter we have seen that recent evolutionary ethical theories have provided plausible, albeit tentative, explanations for the emergence of objective moral norms in human evolution. They have also offered reasonable accounts for our propensity to adhere to such norms. Thus both forms of the moral argument can be weakened by recent naturalistic models, as the former argued that *in principle*, naturalistic frameworks are incompatible with altruism and moral awareness. However, if the evolutionary explanations discussed above (which locate moral facts in human social institutions) are on the right track, they indicate that naturalism is not in principle incompatible with the moral sense and objective moral norms.

7 The Argument from Beauty and the Evolutionary Basis of Aesthetic Experience

The argument from beauty, or the aesthetic argument, denotes a family of arguments for the existence of God that take the beauty of the natural world or works of art (e.g., Bach's cantatas) as evidence for the existence of God. Some authors have articulated this connection between the divine and beauty as obviously true, as something that does not even require an explicit argument at all. For instance, Erazim Kohák (1984, 182) regards aesthetic experiences of nature as an integral part of its religious significance: "In lived experience, in the radical brackets of the embers and the stars, the presence of God is so utterly basic, the one theme never absent from all the many configurations of life's rhythm." In a similar vein, Alvin Plantinga (2000), following Calvin ([1559] 1960), submits that aesthetic experiences of the natural world can elicit the *sensus divinitatis*, the sense of the divine. Such experiences elicit a spontaneous and unmediated sense of God's presence in creation. They include

the marvelous, impressive beauty of the night sky; the timeless crash and roar of the surf that resonates deep within us; the majestic grandeur of the mountains; ... the ancient, brooding presence of the Australian outback; the thunder of a great waterfall. But it isn't only grandeur and majesty that counts; [Calvin] would say the same for the subtle play of sunlight on a field in spring, or the dainty, articulate beauty of a tiny flower, or aspen leaves shimmering and dancing in the breeze. (Plantinga 2000, 174)

Plantinga does not invoke these beautiful and awe-inspiring things in an argument.

It isn't that one beholds the night sky, notes that it is grand, and concludes that there must be such a person as God: an argument like that would be ridiculously weak. ... It is rather that, upon the perception of the night sky or the mountain vista or the tiny flower, these beliefs just arise within us. They are *occasioned* by the circumstances; they are not conclusions from them. The heavens declare the glory of God and the skies proclaim the work of his hands: but not by way of serving as premises for an argument. (Plantinga 2000, 175)

As Plantinga's remark intimates, the aesthetic argument does not follow a classic argumentative structure. An explicit inference from "There is beauty" to "Therefore God exists" seems spurious. Dawkins (2006), one of the few contemporary nontheist authors who devotes some attention to the argument, remarks that the perceived beauty and sublimity of Beethoven's late quartets or Shakespeare's sonnets obviously do not prove the existence of God; if anything, they only establish the existence of Beethoven and Shakespeare (and even so, they do not do so conclusively, as they could have been misattributed). Unlike other arguments for the existence of God, which move from clearly stated premises to conclusions, the argument from beauty has an ineluctably phenomenological component, the experience of beauty and of awe, and its connection to religious experience. In this, it differs from other arguments: even the arguments from religious experience and from miracles can be clearly stipulated and assessed without having had religious experiences or having witnessed miracles for oneself.[1] This is because the connection between religious experience and God, as well as miracles and God, seems obvious. By contrast, the connection between beauty and God's existence is less directly obvious, even if it is convincing when one has an aesthetic experience.

What is the link between aesthetic experience and religious belief? CSR is remarkably silent on this. The prime elicitors of religious belief that CSR identifies are not aesthetic in nature but, rather, are unpleasant stimuli, such as suspicious sounds and sights, for instance, rustling leaves or creaking planks in a dark house. In this chapter, we will draw on evolutionary explanations of aesthetics and art to explore the connection between aesthetic experience and religious belief. Classical theories of aesthetics from the eighteenth century onward identify two aesthetic qualities as primary: beauty and the sublime. Since these two qualities also figure prominently in aesthetic arguments for the existence of God, they provide a suitable guideline for our investigation of the cognitive basis of aesthetic appreciation. We will examine evolutionary and cognitive approaches to beauty and the sublime and look at implications for the aesthetic argument.

Aesthetic Arguments

Currently, the argument from beauty enjoys relatively little philosophical interest and discussion. It is not incorporated in major anthologies of natural theological arguments (e.g., Craig and Moreland 2009). Nevertheless, the argument has ancient roots. Socrates, as recorded in Xenophon's *Memorabilia* ([4th c. BCE] 1997, 55–63), was one of the first to formulate it (see also chapter 1). According to Socrates, the world exhibits objective beauty, a beauty that is best explained by the existence of a creator. He argued that representational artists, such as sculptors, are admirable creators, even though their creations are not living things. Given that living things exhibit teleology, they are brought about by design rather than chance. From this, Socrates concluded that the being that produced living things is more praiseworthy than the most admirable of human artists.

One of the best-developed arguments from beauty can be found in the philosophical theology of Frederick Tennant. Tennant (1930, 89–93) observed that nature is *saturated* with beauty: it is all around us and permeates all of the natural world. Drawing an analogy with human-created works, he judged that manmade artifacts are rarely beautiful, and if they are made without any artistic intent, the probability of them being beautiful is slim indeed. As a case in point, he discussed the automobile—a noisy, smelly, and ugly thing made with only functional concerns in mind. (Nowadays, car manufacturers try to make attractive products, but this underscores Tennant's point: cars only became aesthetically pleasing—to some—when designers started paying attention to aesthetic features of car design.) If the natural world is the outcome of processes that are indifferent to aesthetic value, it is not likely that it exhibits beauty, just like human works made without artistic intent are unlikely to be beautiful. Nevertheless, nature exhibits such uniform and abundant beauty that we should conclude that the assumption that it was created without any regard for beauty must be false. Hence, the cause of nature likely had an active interest in beauty.

Tennant's argument was not meant as a proof for the existence of God. Rather, like the argument from design, with which it shares structural similarities, it is an argument to the best explanation. It thus also shares a weakness with the argument from design (see chapter 4), as both depend on an analogy between the natural world and human artworks and artifacts. It seems reasonable that human-created beauty is usually the result of artistic intent, yet we cannot confirm whether natural beauty likewise requires artistic intent. In response, the defender of the argument from beauty could reply that the burden of proof here lies upon the skeptic, who should provide a reasonable explanation of why nature is saturated with beauty, given that in human works, it is unlikely to arise without prior artistic intent. Remarkably, as Mark Wynn (1997) notes, Tennant's argument does not critically depend on the presupposition that beauty is an objective (i.e., mind-independent) property. For, even if beauty were mind-dependent, we could just ask why it is that nature invariably has the effect on us that we find it beautiful, whereas products of human agency only have this effect when they are designed to be beautiful.

C. S. Lewis's version of the aesthetic argument focuses not on aesthetic experiences themselves but on the human longing for such experiences (e.g., Lewis [1949] 2001, 136–137). This argument from desire takes as a starting point the observation that people have a deep need, a longing that he denoted with the German *Sehnsucht*, which they attempt to fill with aesthetic and pleasurable experiences. Lewis proposed that our aesthetic desires remain unfulfilled and indeed insatiable and, hence, that whatever can satisfy such desires is not from this world. The books and music in which we believe beauty is located do not contain the object of this desire. Rather, it is through them that we glimpse the unreachable object we long for. The sense of the sublime that is present in our experiences of great works of art or natural beauty can be regarded as a manifestation of *Sehnsucht*: it is a sense of being overpowered by one's aesthetic experience and one's longing (and inability) to be part of what one is experiencing: "Beauty has smiled, but not to welcome us; her face was turned in our direction, but not to see us" (Lewis [1949] 2001, 40).

In this respect, the desire for aesthetic experience is quite different from other desires in that the latter can be fulfilled; for example, if one has a desire for ice cream, the desire is fulfilled by eating gelato, but the desire that fills one when enjoying a stunning work of art leaves one unfulfilled. The argument from desire relies on a phenomenological sense of longing, which the reader has to share in order for the argument to get off the ground; if the reader has never experienced this sense of longing, the argument does not get far. The idea that humans have a sense of incompleteness and a desire for the transcendent has been proposed by a number of theologians and philosophers, including Augustine, Pascal, and Kierkegaard (Cottingham 2012). To these authors, these feelings of restlessness and incompleteness derive from a longing for God. As Augustine ([397–398] 1961, 21) wrote, "Our hearts find no peace until they rest in you." Given that all our other desires have objects (e.g., the desire for food can be satiated with a meal, the desire for sex with a suitable partner), Lewis conjectured that *Sehnsucht* also has an object, God. Lewis's argument thus relies on an inference to the best explanation in that the object of a desire figures in the explanation for why we have it. We will use naturalistic explanations of aesthetic desire and of our sense of the sublime to assess the argument from beauty.

Aesthetic Appreciation as Universal Human Behavior

A key premise of the aesthetic argument is that humans perceive the world as saturated with beauty. Lewis made the additional claim that humans have a desire (which he identified as a desire for God) that they attempt to quench with aesthetic experiences. Is there empirical support for this universal human quest for beauty? All known human cultures have at least music, dance, and some form of body adornment, and many others have plastic arts, such as sculpture or painting as well (Brown 1991, 40). This is even the case for societies with very limited material culture, such as the Yaghan, inhabitants of Tierra del Fuego, who traditionally had bark masks, jewelry, and body painting, even though they habitually wore almost no clothes, in spite of living in a mean annual temperature of only 5.5°C/41.9°F (Koppers 1924). Moreover, although not all non-Western languages may have words to denote art, they do have indigenous terms that capture aspects of the Western concept of art, such as skill or beauty (Van Damme 1997). In almost all cultures, including small-scale societies that do not have much division of labor or social stratification, there is a recognition of artistic expertise: some people are widely acknowledged within their community as being more skilled in producing aesthetically pleasing artworks, be they musical compositions, skillfully told stories, or masterly carved sculptures (Anderson 1989). This indicates a cross-cultural concern with beauty and the arts.

Art making and aesthetic appreciation are universal at the cultural level, but are they also universal at the individual level? This seems likely. Even people who never visit an art gallery or a concert have personal aesthetic tastes, for instance, for particular musical styles and artists, or in furnishing and decorating their homes and gardens. Most untrained people are fairly skillful reproducers of music as well: they can sing tunes relatively accurately, and they do not make many errors in pitch and tempo (Dalla Bella, Giguère, and Peretz 2007). Art making and aesthetic sensitivity emerge early and spontaneously in child development, as is evident in an early disposition to draw, sing, dance, and play word games (Dissanayake 2000). Children require little, if any, prompting to engage in such behaviors, as compared to, for instance, reading and writing, which require long and explicit instruction. Although children can be taught to be better artists, the propensity to create aesthetically pleasing objects and performances develops very early.

Archaeological evidence for aesthetic concern and art making can be found in *Homo sapiens* populations across the Old World. The earliest convincing examples are body decoration: beads made of perforated seashells from Israel and Algeria, dated to 135,000–100,000 years ago (Vanhaeren et al. 2006). More elaborate forms of body decoration arise somewhat later, with beads made of finely polished circular fragments of ostrich eggshell from Enkapune Ya Muto, Kenya, dated to 50,000 years ago (Ambrose 1998). The earliest figurative sculptures are about 40,000 years old. They are small mammoth ivory figurines from Swabia, southwestern Germany, that represent animals, therianthropes (e.g., the so-called lion man) and humans (Conard 2003). The oldest securely dated figurative paintings have been found in Europe (Chauvet Cave, France, 32,600 years ago and El Castillo, Spain, 40,800 years ago); they show hand stencils and representations of carnivores and herbivores (Valladas 2003; Pike et al. 2012). Australian rock paintings in the Kimberley region (northwestern Australia) are probably more than 40,000 years old. Although direct dating of these paintings is difficult, several of them show accurate, naturalistic depictions of extinct animals such as marsupial lions, giant kangaroos, and giant flying foxes, species that died out between 46,000 and 40,000 years ago (Akerman and Willing 2009).[2] By about 40,000 years ago, humans also made music, as is clear from fragments of ancient flutes made of perforated bird bones found in southwest Germany (Conard, Malina, and Münzel 2009). Evidence for dance can only be indirectly inferred, from Magdalenian representations of dancers, such as the dancing "shaman" (half-human, half-deer being) in the Grotte des Trois-Frères, France, dating to about 13,000 years ago (Clottes, Menu, and Walter 1990).

Some archeologists suggest that aesthetic sensitivity predates the emergence of our species. From about 1.5 million years ago, hand axes appear in the archeological record. These highly symmetrical tools are about the size of a hand, shaped like a wedge or a teardrop, often finely finished beyond purely utilitarian concerns. Remarkably, an analysis of more than 800 hand axes, found at 148 sites across the Old World shows that their proportions (breadth/length relationship) conform approximately to the golden ratio (Pope, Russel, and Watson 2006). Taken together, the cross-cultural universality of art, the spontaneous development of artistic behavior in children, and its antiquity in the archeological record indicate that a sensitivity and concern for beauty, and a propensity to create beautiful things are stable parts of human cognition and behavior.

Evolutionary Aesthetics

The second step in the aesthetic argument connects the experience of beauty with God, where God is regarded as the best explanation for this ubiquitous perception and quest for beauty. To evaluate this part of the argument, we need to look at alternative, naturalistic explanations of aesthetic appreciation. Eighteenth-century aesthetic theories (e.g., Kant [1790] 1987) considered aesthetic appraisal as distinct from other kinds of evaluation. According to such theories, aesthetic judgments are characterized by a *disinterested attitude*: when we evaluate the tastiness of an apple or the functionality of a chair, we are concerned with utility and function. By contrast, evaluating the delicacy of a flower or the boldness of a painting does not involve such practical considerations.

Experimental aesthetics has challenged this radical division between everyday and aesthetic judgments, especially the notion of disinterestedness. Biological explanations of aesthetic appreciation date back to the nineteenth century. Experimental aesthetics (see Aiken 1998 for an overview) was in fact among the earliest domains of experimental psychological investigation. Experimental aestheticians examine what mediates our judgment of beauty and its opposites, dullness, ugliness, and blandness, and why we make such judgments. In a psychological sense, aesthetic evaluations are sensory and qualitative appreciations of natural or manmade things. Such appraisals allow one to experience a subjective sense of positive affect (e.g., pleasure, awe) or negative affect (e.g., disgust, sadness), what psychologists collectively refer to as aesthetic experiences. Evolutionary aesthetics is the branch of experimental aesthetics that is concerned with the *why* of aesthetic experiences and judgments, that is, their ultimate level of explanation. As such, it can provide us with naturalistic explanations for the human quest for, interest in, and perception of beauty. In what follows, we will look at four types of evolutionary aesthetic theories of human aesthetic experience and judgment: sexual selection theory, evolved sensory bias theory, the biophilia hypothesis, and the prototype approach to awe. We will examine whether they provide a plausible naturalistic explanation for why humans perceive the world as saturated with beauty.

Beauty and Sexual Selection

Many species of mammals, fish, birds, reptiles, and insects have decorative features such as long feathers, elaborate antlers, bright colors, "carried ... to a wonderful extreme" (Darwin 1871b, 78). They engage in behaviors such as singing (e.g., passerine birds, gibbons) or decorating elaborate constructions with brightly colored flowers and other objects (e.g., bowerbirds). Initially, Darwin and Wallace were puzzled by this. A peacock's tail does not contribute to its owner's ability to survive—if anything, the bird is less able to escape predators and more susceptible to parasites. Early natural selection theory held that only traits that contribute to an organism's capacity to survive are selected. Animal ornaments thus presented an anomaly to this theory.

To explain why animals have beautiful ornaments or why they engage in behaviors that are aesthetically pleasing, Darwin came up with sexual selection, appealing to aesthetic judgment as a driving force. Animals (usually females) prefer mates that are beautiful (e.g., have bright colors or sing elaborately). As a result, such traits will spread throughout the population.

Sense of beauty.—This sense has been declared to be peculiar to man. But when we behold male birds elaborately displaying their plumes and splendid colours before the females, whilst other birds not thus decorated make no such display, it is impossible to doubt that the females admire the beauty of their male partners. ... If female birds had been incapable of appreciating the beautiful colours, the ornaments, and voices of their male partners, all the labour and anxiety exhibited by them in displaying their charms before the females would have been thrown away; and this it is impossible to admit. (Darwin 1871a, 63)

This attribution of an aesthetic sense to animals was not meant as a form of anthropomorphism but can be situated within Darwin's broader project of looking for continuity between humans and other animals. In this way, he aimed to uncover the evolutionary origins and functions of human faculties. Darwin's view of aesthetics was strongly influenced by eighteenth-century philosophical theory, in that he thought beauty was nonutilitarian: beautiful traits in animals served no other purpose than being attractive (Menninghaus 2009). He distinguished sexual selection, which favors aesthetic traits, from natural selection, which selects for functional, useful traits.

While sexual selection theory may explain animal ornaments, it does not explain aesthetic preferences. Why do animals prefer mates that are brightly colored or sing elaborate songs rather than mates with dull colors and monotonous songs? If aesthetic preferences are purely arbitrary, why do females seem to go for extravaganza rather than discretion? Darwin explained this by appeal to a coevolution of taste and appearance: "The male Argus pheasant acquired his beauty gradually, through the females having preferred during many generations, the more highly ornamented males; the aesthetic capacity of the females having been advanced through exercise or habit in the same manner as our own taste is gradually improved" (Darwin 1871b, 401).

Wallace was unconvinced by Darwin's appeal to arbitrary aesthetic preferences. While he did not deny the possibility of evolution by mate choice, he thought that aesthetic preference was not purely arbitrary but that ornaments correlated with quality. For example, to explain the bright colors in some bird species, Wallace (1877, 401) mused, "If, however, heightened colour is correlated with health and vigour, and these healthy and vigorous birds provide best for their young, and leave offspring which, being equally healthy and vigorous, can best provide for themselves, then natural selection becomes a preserver and intensifier of colour."

Interestingly, Wallace's view, and not Darwin's, currently prevails in evolutionary biology (Prum 2012). The leading neo-Wallacean hypothesis to explain sexual ornaments is Amotz Zahavi's (1975) handicap principle. Zahavi argues that bright, loud, and colorful ornaments are a handicap: they make their owners less mobile or more visible and more vulnerable to predation. Only high-quality individuals are able to bear the handicap and yet survive to adulthood. Thus, they provide a good proxy to the quality of potential mates. Signals that are costly are typically hard to fake and can be preferred because they are reliable. If females systematically go for the largest handicap, traits such as long tails and bright colors will get more pronounced over the next generations. These neo-Wallacean explanations are limited in scope: they can explain, through a mechanism of coevolution between traits and a preference for them, why we make aesthetic judgments about and derive aesthetic enjoyment of potential mates and things that resemble them (such as sculptures or paintings representing people of the opposite sex), but they do not explain why we derive aesthetic pleasure from delicate cherry blossoms or stormy seascapes. The evolutionary psychologist Geoffrey Miller (2000) puts forward a broad neo-Wallacean theory to explain why we enjoy and produce art. Human artworks are like the bowers of bowerbirds: they are a form of courtship display. Sculpting, painting, music making, and storytelling require considerable energy and time. Their costliness provides a reliable signal of the fitness of the art-producing person—in Miller's view, primarily an art-producing male in his prime. Just like a lush but burdensome tail in a male bird of paradise signals its owner's qualities that allow it to live with such a handicap, artworks honestly display the artist's qualities as a mate. Miller does not provide an account for why we find *nature* saturated with beauty, a key premise of the aesthetic argument. Indeed, he predicts that we prize mainly man-made objects and performances. Nevertheless, aesthetic appreciation of nature is cross-culturally ubiquitous, such as of the ephemeral beauty of cherry blossoms delicately dancing in the breeze (Japan), wide vistas of mountains (China), or a tempestuous sea (Europe). Sexual selection theory thus does not provide an adequate naturalistic explanation of our appreciation of nature.

Aesthetic Appreciation and Evolved Sensory Biases

As an alternative to sexual selection, some authors (e.g., Barry 2006) have proposed that aesthetic judgments are evaluative mechanisms that have evolved to deal more effectively with our surroundings. These *sensory bias theories* hold that aesthetic experiences draw on perceptual and motivational neural structures involved in everyday life. Given that our senses are constantly bombarded by impressions, the nervous system needs to prioritize some cues over others. According to Ann Barry, aesthetic preferences find their origin in the brain's reward system, which guides attention to relevant perceptual input that is likely to yield information important for survival and reproduction. Most proponents of sensory bias theories are neuroscientists who study the neural correlates of our appreciation of art and other stimuli. Their findings can be summarized as follows: assessing artworks and other beautiful stimuli taps into normal propensities and biases of our sensory systems; aesthetic appraisals are primarily emotional and driven by reward- and motivation-related areas of the brain. In accordance with the redeployment hypothesis (see chapter 2), there is not one single area in the brain that deals with aesthetic appreciation but rather a network of areas involved with sensory perception, motivation, and reward (see De Smedt and De Cruz 2010 for an overview).

One prediction of sensory bias theories is that successful art homes in on biases and predispositions of human perception. Artists indeed tend to represent and emphasize those features that our visual and auditory systems are attuned to. To take an extreme example, the strong horizontal and vertical lines in the work of artists like Piet Mondriaan and Kazimir Malevich activate orientation-selective cells in the primary visual cortex (area V1) that respond selectively to dots and straight lines, especially to horizontal and vertical ones. Conversely, artists tend to ignore or underplay features that are less important to human perception, such as shadows. Humans are notably bad at correctly interpreting shadows, and, unsurprisingly, these are absent or incorrectly rendered in many paintings and other pictorial work (Cavanagh 2005). This intimate connection between the function of art and the function of the brain led Semir Zeki (1999, 10) to quip that artists are in a sense neuroscientists since art, in order to be successful, must appeal to human perceptual, conceptual, and motivational systems. In other words, art appeals to us because it exaggerates or appropriates features that human perception is attuned to.

Our aesthetic response to artworks is primarily reward-driven and emotional. Listening to music that gives us chills or looking at paintings we find beautiful activates reward- and motivation-related brain areas (Blood and Zatorre 2001; Vartanian and Goel 2004). A recent meta-analysis of ninety-three papers in neuroaesthetics found that the anterior insula (bilateral, but mostly the right) is strongly involved in aesthetic appreciation (Brown et al.

2011). The right anterior insula is a part of the limbic system that is consistently involved in mediating pain, sadness, cravings for food, recreational drugs, and providing an emotionally relevant context for perceptual experiences. More generally, this area of the brain is involved in the representation of the subjective self based on bodily awareness (such as taste, pain, longing), which corroborates the hypothesis that aesthetic appreciation is not a disinterested affair but rather emotional and reward-driven. However, sensory bias theories do not explain why many highly acclaimed works of visual art are hardly eye candy, such as Pablo Picasso's stark, uncompromising *Guernica* (1937). It predicts that artworks that exaggerate features that are salient to our processing of everyday stimuli will be more successful. This prediction is partially supported by the evolution of the work of individual artists, such as Paul Klee, Piet Mondriaan, and Henri Matisse, where one can observe the increasing use of strong lines, vivid colors, and bold contrasts. Nevertheless, abstract art does not appeal to a broad audience in a way we would expect if sensory biases were the main guide to aesthetic appreciation. Moreover, sensory bias theory, by its focus on early perceptual processing, is silent on our appreciation of nature. It seems therefore that sensory bias theories do not capture the full range of our actual aesthetic experiences.

The Biophilia Hypothesis

E. O. Wilson's biophilia hypothesis (1984) submits that humans have an innate, evolved disposition to be attracted to and aesthetically respond to nature. It is currently the best-developed naturalistic theory to explain our sense that nature is saturated with beauty. For the largest part of our evolutionary history, until a few hundred years ago with the advent of the Industrial Revolution, human biology has been embedded in natural environments, first as hunter-gatherers (from about 2.5 million years to about 10,000 years ago), then as horticulturalists and farmers (from about 10,000 years ago). It seems plausible that an affinity with natural environments provided a survival advantage: finding water sources, prey animals, and edible plants was—and in some remnant hunter-gatherer populations still is—a matter of life and death. This evolutionary scenario predicts a continued responsiveness to nature in humans who live in heavily urbanized environments. Even in industrialized countries, where contact with nature is relatively sparse, depictions of animals and plants have not diminished in popularity: drawings by young Western children teem with flowers, trees, quadrupeds, birds, and fish.

The biophilia hypothesis derives support from a variety of converging lines of evidence. Studies indicate that proximity to nature increases well-being, including health, happiness, and life satisfaction. Many of these studies are not properly controlled but rely on observations in real-life settings. While they do not prove the biophilia hypothesis unequivocally, they are often taken to support it through a consilience of inductions. For example, patients who were recovering from surgery in rooms looking out on a small park with deciduous trees healed faster than patients who were looking out on a brick wall (Ulrich 1984). In a natural experiment in a large state prison in southern Michigan, half the prisoners were assigned to cells that had a view of farmlands and forests, with the other half looking out on the concrete prison courtyard. Prisoners confined in cells overlooking the courtyard had a 24 percent higher frequency of sick call visits (Moore 1981). Across cultures, people find natural environments more aesthetically appealing than urban surroundings (reviewed in Ulrich 1993), a remarkable fact given that humans built their urban environments to live in. The proximity of nature has a strong influence on real estate prices. Apartments that are otherwise similar in quality find buyers willing to pay between 8 percent and 60 percent more if they provide a view of a park, a lake, or the ocean. This preference for a home with a scenic view is cross-culturally robust, attested in the Netherlands, the United States, Hong Kong, New Zealand, and mainland China (Jim and Chen 2009). Our valuing of nature (including wildlife) is also reflected in the popularity of zoos, parks, and nature reserves, which attract more visitors per year than all sports events combined (Frumkin 2001).

While these empirical findings support the biophilia hypothesis, it remains problematic. One of its difficulties is testability: it is easy to pick evidence in favor of biophilia, but one can equally give evidence against it. After all, people seem to be relatively unconcerned with destroying large swaths of pristine nature, such as the equatorial rainforest, and most of us feel discomfort or fear for particular biological stimuli, such as wasps or spiders. To be more testable, the biophilia hypothesis has been refined: humans have an evolved preference not for nature in general but for some environments over others. For a large part of their evolutionary history, our East African hominid ancestors have lived in savannah-like landscapes, large areas of grass in close proximity to gallery forests and water. Water sources not only provided drink, but were gathering places for the herbivorous animals that were the main source of food for hominids that first scavenged and later hunted them. A focused version of the biophilia hypothesis holds that humans have an evolved preference for landscapes that resemble this park-like environment. It has found good empirical support. Even people who have not grown up in savannah-like environments show an aesthetic preference for such landscapes: inhabitants of the Nigerian rainforest belt as well as American city dwellers aesthetically prefer photographs of savannah landscapes to pictures of rainforest, desert, deciduous woodland, and pine forest (Falk and Balling 2010). This preference for savannah-like landscapes emerges in young children from diverse ecological backgrounds (Orians and Heerwagen 1992).

However, this more developed form of the biophilia hypothesis fails to account for our aesthetic preferences of scenery that does not adhere to the savannah schema. Consider our sense of awe and aesthetic enjoyment of the night sky. Why this aesthetic preference, given that stars did not play an important role in our evolutionary history? We are, after all, not indigo buntings (*Passerina cyanea*), a species of bird that orients itself using star patterns during its biannual migrations. Indigo buntings pay attention to the night sky even as nestlings, and if we similarly relied on the stars, our fascination with the night sky would be more easily explained. In a similar vein, Plantinga's majestic mountains or brooding outback are not environments that are conducive to human flourishing. One way to explain this aesthetic response is to appeal to culture. The enjoyment of the mountains and the sea is a relatively recent phenomenon in Western culture—these places were perceived as threatening and hostile until well into the nineteenth century. It was only in the course of the late eighteenth century that people started to visit the beach for recreational purposes (Corbin 1988). But while such explanations indicate the malleability of our aesthetic responses by culture, they do not really explain why we have those responses in the first place. As Sperber (1996) points out, one cannot explain culture, in this case, a cultural preference for landscapes that did not play a role in our evolutionary history, by invoking culture. Surveying the evidence on evolved preferences for landscapes, Stephen Davies (2012, 98) concludes: "It is far from clear how this [culture dependence] is consistent with the thesis of an environment of evolutionary adaptedness or the landscape aesthetic that is most widely espoused by evolutionary psychologists."

An alternative route might be to say that while we find savannah-like landscapes beautiful, we find mountains, the night sky, and the desert sublime. Since the biophilia hypothesis focuses on subjective judgments of beauty and feelings of well-being, it leaves open the possibility that different aesthetic judgments (such as a sense of the sublime) underlie our experience of hostile and threatening landscapes. As far as we are aware, there is only one naturalistic theory that attempts to explain the sense of the sublime. We will examine this in the next section.

Cognitive and Evolutionary Explanations of the Sense of the Sublime

The term *sublime* refers to qualities that inspire a sense of awe and wonder with a hint of challenge or danger. Sweeping mountainous landscapes, Niagara Falls, and the Northern Lights are classic examples. Awe (the emotion most associated with the experience of the sublime) has received some attention in aesthetics (e.g., Burke 1757), but there is surprisingly little psychological work on its cognitive foundations. In keeping with this general trend, evolutionary aesthetics has examined mainly beauty but virtually ignored how perception of the sublime contributes to aesthetic experience. This is perhaps because awe is not unique to aesthetic experience, as it can be triggered by a variety of stimuli, including mathematical equations, scientific theories, charismatic leaders, heroic acts, tornadoes, and transformative life experiences such as childbirth.

The *prototype approach to awe* (Keltner and Haidt 2003) is at present the best-developed psychological framework for the study of awe. Dacher Keltner and Jonathan Haidt propose that two features lie at the basis of prototypical cases of awe: vastness and the need for accommodation. *Vastness* is the experience of something being much larger than oneself. This can be purely a matter of size, as the experience of the starry night sky or of palatial ballrooms, but also social size, such as prestige, authority, or fame. As suggested by Kant ([1790] 1987), vastness can also be dynamic, when it expresses power or strength, for example, the might of a stormy sea or a hurricane. It can also be implied in other ways; for example, one may encounter vastness in the grandeur of a scientific theory, not because it is literally long, but because of its scope and predictive power, as can be seen in the closing sentence of Darwin's *Origin of Species*:

There is *grandeur* in this view of life, with its several powers, having been originally breathed into a few forms or into one; and that, whilst this planet has gone cycling on according to the fixed law of gravity, from so simple a beginning endless forms most beautiful and most wonderful have been, and are being, evolved. (Darwin 1859, 490, emphasis added)

The *need for accommodation* is a psychological process by which one attempts but fails to assimilate a new experience. Reactions to disorienting and frightening experiences, such as Paul's vision of Christ on the road to Damascus (Acts 3:3–6) or Arjuna's vision of the world given to him by Kṛṣṇa (*Bhagavadgītā*, 11:9–46), are examples of this. Failed attempts at accommodation can also occur with less frightening stimuli, such as seeing a stunning landscape like the Grand Canyon. Awe can be "flavored" by diverse other emotions such as fear (as when one is in awe of hurricanes demolishing buildings) or beauty (being overwhelmed by a Renaissance motet).

Keltner and Haidt look into our primate ancestry for clues to explain features of awe. They propose that the prototypical case of awe is that of a subordinate in the presence of a dominant individual. Like most other apes, humans have complex dominance hierarchies.[3] Emotions play an important role in the maintenance of dominance hierarchies and may have adaptive value. For instance, a sense of pride might give one the motivation to obtain a dominant position and defend it against rivals. By contrast, individuals who are not able to attain a dominant position might do better if they defer to higher-quality group members. This could lead to the evolution of emotions that discourage attempts to overturn the social hierarchy, such as awe. Perceived vastness, failure to accommodate, and the willingness to surrender are feelings that may accomplish this.

Louise Sundararajan (2002) expands Keltner and Haidt's model by adding the component of *self-reflexivity*. Awe frequently brings about a self-reflective attitude: we are having the experience and also perceive ourselves as living it. This form of self-reflexivity is frequently encountered in the writings of mystics, who express a willingness to be wholly absorbed by or surrender to this experience. This is what Lewis denotes as *Sehnsucht* or what Rudolph Otto (1923, 21–22) calls "self-depreciation." Remarkably, this self-reflexivity results in a sense of humility, whereas other positive emotions such as happiness lead to an increased focus on oneself and a more positive appraisal of oneself. Self-reports of recent aesthetic experiences of natural beauty indicate that participants are likely to feel a sense of smallness or personal insignificance, a decreased awareness of day-to-day concerns, a sense of something greater than oneself, and a desire for the experience to continue (Shiota, Keltner, and Mossman 2007).

Keltner and Haidt's account of awe fails to explain why not just powerful individuals but also divergent stimuli such as landscapes, the starry night sky, or a choral work by Palestrina elicit this emotion. Indeed empirical research indicates that other human beings are not the primary elicitors of awe, as they predict. For instance, Caldwell-Harris et al. (2011) found that the three most important elicitors of awe in Christians, Buddhists, and atheists are nature (54 percent), science (30 percent), and art, especially music (12 percent). Shiota et al. (2007) found that the most common elicitors of awe are nature (27 percent) and art, especially music (20 percent). The prototypical example of an awe-inducing stimulus in both studies is a panoramic scene of natural beauty. If awe evolved to modulate our behavior adaptively in response to powerful others, why is its main elicitor a nonsocial entity? The current best scientific explanation for awe does not explain why natural beauty is its most important elicitor.

Linking Aesthetic Properties with God's Existence

Arguments from beauty, while lacking a strong argumentative structure, propose that aesthetic experience, or a desire for this experience to continue, is linked to the existence of God. While, as we have seen, there are a number of naturalistic explanations for our subjective sense of beauty and the sublime, these do not provide a good explanation for why we experience nature as saturated with beauty. Can a proponent of the aesthetic argument maintain that God is a plausible explanation for aesthetic experience and judgment? Failure to come up with an alternative hypothesis to explain a given phenomenon can provide nonempirical evidence in favor of a hypothesis (Dawid, Hartmann, and Sprenger in press). If God's creative work is a viable explanation for why we find nature saturated with beauty, and there is no alternative naturalistic explanation for this, in spite of significant efforts to devise one, this constitutes defeasible evidence for God's existence. It is defeasible because such an explanation may be offered in the future (for analogous cases, see chapter 4 on the evolution of language and chapter 6 on naturalistic explanations of the moral sense).

To do this successfully, defenders of the aesthetic argument need to establish why God would be a good explanation for our experience of beauty and the sublime. While it may seem evident to a theist that God would want to create beauty, a nontheist may simply experience beauty as the sense of being moved by nature and its properties, or, in the case of art, being moved by artworks. For instance, Noël Carroll (1993) regards aesthetic experience of nature as a visceral form of appreciating nature for what it is, not as some religious sentiment. The aesthetic argument should make explicit how aesthetic experience is explained by theism.

There are at least two ways to link God and beauty: by appeal to design, and by regarding beauty as an element of religious experience. As we pointed out earlier, many versions of the aesthetic argument (e.g., Socrates, Tennant) make a connection between aesthetic properties and design. They infer the existence of a creator from the perceived beauty and harmony of creation. This strategy is commonly found in Patristic and early medieval writings, including Basil of Caesarea and Augustine. An alternative view is to see beauty as a form of religious experience: to experience beauty is to experience God. In this interpretation, beauty somehow reflects God's nature. In Christianity, the Eastern Orthodox tradition has maintained an intimate connection between beauty and holiness, which it inherited from the Patristic tradition. Accordingly, beauty allows humans to share in God's nature (Viladesau 2008).

Beauty as a Result of Design Intentions
Aesthetic arguments can consider why God would design humans in such a way that they have aesthetic experiences. For example, Polkinghorne (1998, 82) argues that our perception of beauty in the world can be explained as a sharing in its maker's delight in creation. Karl Rahner (1982) thought that some artworks, for example, symphonies by Bruckner or paintings by Rembrandt, are inspired by God to say something about what humans are that cannot be expressed in words. Together with other human cognitive features, such as our moral intuitions (see chapter 6) or our knowledge of the empirical world, appreciating beauty provides us with an insight into God's mind.

The *artistic design stance* can provide a useful analog to explore this idea further. In chapters 2 and 4, we saw that humans have a design stance by which they infer the intentions of the makers of artifacts. Applied to artworks, the design stance allows viewers to infer the complex causal history of artworks, taking into account their art historical context. Often, this is a mediate activity, as when we infer the intentions of the artist based on art historical considerations. This is what Bullot and Reber (2013) have termed "taking the artistic design stance." For example, a dog in a medieval painting, such as in Jan Van Eyck's portrayal of the Arnolfini bridal couple, can be interpreted as a symbol for fidelity in marriage because dogs were a symbol of fidelity for medieval artists. But recognizing design also relies on non-inferential cognitive processes. For instance, when reading Jane Austen's novels, we gain insight into her mischievous delight in creating morally poignant characters and situations. Such insight does not follow explicit inferences about Regency culture but, rather, relies on the direct experience of reading her work, encountering the characters, and vicariously living the situations she depicts.

Keith Lehrer (2006) asserts that we gain ineffable and immediate knowledge of an artwork by directly interacting with it. A verbal description of its content still leaves out something essential of what that work is like. A detailed account of, say, Vincent van Gogh's intentions when painting *Starry Night* (1889), based on ego documents and art historical information, does not impart all of his intentions. This knowledge can only be gained by perceiving the artwork itself, for example, design intentions evident in the bold color contrasts and the whirling brushstrokes. Conversely, even if all information on van Gogh's life, work, and cultural context were destroyed, we would still know something about these design intentions as long as we have perceptual access to his oeuvre (De Smedt and De Cruz 2013a). It is in this context that we can understand Polkinghorne's aesthetic argument: enjoying the natural world provides an ineffable knowledge of God's artistic design intentions. By endowing humans (and presumably also other animals) with a sense of beauty, God grants insight in his relationship with creation (for a further exploration of these ideas, see De Smedt and De Cruz 2013b). As we saw in chapter 4, adopting the design stance critically depends on background information. For a theist, the picture of God wanting to share his design intentions through beauty may seem plausible, but it may not convince those who do not regard the natural world as a created entity.

Aesthetic Experience as Religious Experience

A second way to connect aesthetics and God is by regarding aesthetic experience as a subspecies of religious experience. God reveals something of his nature through the beauty and sublimity of his creation. This view has been developed extensively by the Syrian mystic Pseudo-Dionysius the Areopagite (5th–6th c.), who wrote that God's beautiful nature shines through in creation in the beauty, harmony, and splendor of all things. Twentieth-century theologians such as Karl Barth and Hans Urs von Balthasar have further elaborated this idea of divine beauty (see Viladesau 2008 for an overview). Thus conceived, the aesthetic argument is not an explicit argument anymore but makes an externalist case: forming theistic belief through aesthetic experience is justified because there is a proper connection between God's existence and aesthetic properties in the world. Plantinga (2000), a contemporary defender of this view, argues that aesthetic experiences of nature can elicit the *sensus divinitatis*. These experiences draw forth belief in God in a non-inferential manner: he has designed our cognitive faculties in such a way that they will form theistic belief under a broad range of circumstances, for instance, when confronted with beauty or with one's moral sense. To Plantinga, the proper functioning of our cognitive faculties justifies belief in God formed through aesthetic experience.[4]

Here, the aesthetic quality of the sublime plays a prominent role. Empirical studies have shown that the experience of the sublime increases religious belief. Valdesolo and Graham (2014) showed participants a video montage of nature clips, composed primarily of grand, sweeping views of plains, mountains, and canyons. The control group saw nature videos with animals involved in amusing antics. The participants who witnessed the awe-inducing nature scenes reported higher belief in a God who providentially controls the world than the subjects who saw the funny videos. Saroglou, Buxant, and Tilquin (2008) used a similar method and found that people who saw natural beauty reported higher levels of spirituality.

The relationship between the sublime and religion is thus experiential rather than the result of a premise-conclusion style argument. The Jewish theologian Abraham Heschel did not want to prove the existence of God through our awe of nature but took the existence of God as a given and regarded awe and wonder at natural beauty (not only vast things, but also diminutive things such as snowflakes) as an essential element of religious sensibility. To Heschel ([1955] 2009, 74), awe "is more than an emotion; it is a way of understanding." This understanding is like perception; it provides immediate insight:

> It is not by analogy or inference that we become aware of it [transcendence]. It is rather sensed as something immediately given, logically and psychologically prior to judgment ... a universal insight into an objective aspect of reality, of which all men are at all times capable. (Heschel 1965, 77)

A cultivated sense of awe and wonderment at nature

> enables us to perceive in the world intimations of the divine, to sense in small things the beginning of infinite significance, to sense the ultimate in the common and the simple; to feel in the rush of the passing the stillness of the eternal. (Heschel [1955] 2009, 75)

Clifton Edwards (2012) regards an aesthetic, religious sensibility of the world as a form of practice, an outlook that is afforded through a sensibility to natural beauty. Just like a scientist can gain access to a sphere of knowledge through practice (e.g., an engineer can easily discern the supporting components of a structure), where she learns to perceive things that would otherwise escape her, a religious person may get knowledge of God through acquiring *beauty skills*. For Plantinga, Heschel, and Edwards, aesthetic experience provides a direct, non-inferential, and experiential ground for religious belief. In this way, a religious believer need not provide any direct links between aesthetic experience and the existence of God, just like someone who believes "this tree exists" based on perception need not explain how her perception connects to her belief. As long as this experience is connected to God's existence in the right way (for instance, because God instills this belief in us through aesthetic experience), this belief is justified.

Aesthetic Experience and Religious Fictionalism

The argument from beauty infers the existence of God from the experience of beauty. However, as with the other arguments we have encountered in this book, this inference depends on prior beliefs, in particular, the probability one assigns to the existence of God. Some nontheists are attracted by the aesthetic dimension of religious practice without coming to religious belief. So the fact that one experiences beauty, in particular beauty within religious contexts (the poetry of the psalms, communal hymn singing, arousing artworks, gripping sermons) does not always lead to an inference to the existence of God. *Religious fictionalism* holds that a person can coherently engage in practices such as prayer and religious discourse without endorsing a realist position about God's existence (Eshleman 2005). Following in-depth interviews with Christian atheists, Brian Mountford (2011) found that aesthetic experience in religious painting, music, and literature was one of the main reasons for remaining observant.

The philosopher of religion Howard Wettstein (2012) provides a detailed philosophical defense of religious fictionalism, arguing that the moral and aesthetic dimensions of a religious lifestyle can be enjoyed by nontheists. Drawing on his own tradition, Judaism, Wettstein looks at awe as a form of practice rather than a doxastic attitude. Awe, rather than belief, is the cardinal attitude of observant Jews: *yirat hashem*, the awe of God, is almost a synonym of religion, and a religious person is not merely a believer, but *yare hashem*, he who stands in awe of God. The theist has a sensibility or aptitude, which is nurtured and trained through practices such as prayer, study of the Talmud, ritual, and *berakhot*, regular blessings uttered at various occasions. Wettstein makes the radical claim that one can be in awe of God without actually believing in God: one can conduct rituals and voice blessings at numerous moments of the day, actions that nurture a sense of awe at the world, a sensibility to morality and to beauty, without believing that God exists.

The philosopher of religion Eleonore Stump (1997) is skeptical of this, arguing that belief in God is an essential element of religious faith. The aesthetic element of the religious lifestyle is greatly enhanced by doxastic faith: it will be a lot easier and effective to maintain motivation to study the Talmud if one believes God exists than if one simply regards this study as a way to train one's mind to have a particular sensibility about the world. Besides, if one does not believe in God, why accord any special status to the Talmud or the Gospels? More generally, while faith is more than belief, it does have doxastic elements, and it seems doubtful that one can simultaneously have faith that p while believing that not p (Pouivet 2011). Taking away the doxastic dimension alters the experience of religious art:

Consider, by way of comparison, Homer's description of Apollo shooting plague-causing arrows at the helpless Greek army because Agamemnon has dishonored Apollo's priest. The passage is poetically powerful; it catches one's attention and remains in one's memory. But it's hard to imagine its stimulating religious awe in us or giving us a sense of the holy. And that's precisely because we think that there is no supernatural being such as Apollo, that no deity causes diseases by shooting arrows into people, and that there are perfectly good naturalistic explanations of diseases which have nothing to do with Apollo's seeking revenge for being dishonored. Just because we have such naturalistic views, though the literary beauty of the passage may strike us, it hardly moves us to religious awe. (Stump 1997, 285)

Similarly, many people are struck by Western religious music, such as Bach's *Matthew Passion* or the Gregorian mass for the dead. The promulgation of Christian beliefs was an important, even central element of such musical traditions. Can a nontheist have the same aesthetic experiences listening to them as someone who endorses these beliefs? Neill and Ridley (2010) maintain that an atheist can share such experiences through fictionalism: one enjoys the Christian elements as purely fictional, just like one can be moved by Anna Karenina's suicide, full well knowing she never existed. Robin Le Poidevin (1996, 118–121) also draws attention to this parallel between art as fiction and religion as fiction: just like we can become emotionally involved with fiction through some form of make-believe (as in Anna Karenina's case), the fictionalist holds that we can make-believe that there is a God. Even if the religious beliefs are true in a realist sense, religious art still requires imagination as well, for instance, in the way Bach translated Matthew's passion into music, or in the way a group of Balinese dancers act out scenes from the *Mahābhārata*.

This equivocation of imagination with make-believe is problematic. Even if Christ's death and resurrection or the battle between the Kaurava and Pāṇḍavaḥ princes are compelling from a narrative point of view, a realist position (i.e., a belief that these events occurred) still places one in a better epistemic position to explain the religious stance that believers adopt toward these narratives. Differently put, if the beauty of religious art or scripture compels individuals to adopt a religious lifestyle, it is much more straightforward to say that they do so because they nurture or foster religious beliefs than to say that they do so because these practices allow them to engage in religious make-believe (Pouivet 2011). Moreover, fictionalism seems to work better for beauty experienced in liturgical context and in religious art than as a stance to make sense of the wonder felt for natural beauty. How could a fictionalist interpret the awe one experiences on a majestic mountain? Wettstein would reply: this is awe for God. But here, it seems more straightforward for the naturalist to interpret this as awe for nature (as Carroll 1993 insists) rather than awe for a God one does not believe in. It is thus more difficult for fictionalists to interpret aesthetic experiences religiously than it is for realists. Aesthetic experiences may not reliably lead to belief in the existence of God, but theism still seems the most compelling attitude to adopt in the face of the aesthetic dimensions of religious art and practice, especially when experiencing awe for the beauty of nature.

Summary

The aesthetic argument for the existence of God relies on the observation that humans universally have a concern and appreciation for beauty, and that this appreciation is directed to natural beauty as well as created artworks. This chapter has reviewed work from evolutionary aesthetics that indicates that humans are indeed sensitive to beauty, especially natural beauty. There is at present no satisfactory naturalistic explanation for why humans value natural beauty that does not conform to their evolved tastes. Hence, the proponent of the aesthetic argument can hold that God is currently the best explanation for this sense of beauty. However, it remains hard to link God and beauty in such a way that the existence of God is not already presupposed. The argument from beauty is perhaps most compelling to those already convinced that theism can plausibly explain beauty as a design feature of the world or as a form of religious experience.

As we saw in chapter 1, natural theology not only consists of providing arguments for or against the existence of God, but is also the project of embedding experiences of the world, such as the experience of beauty, within a theistic framework. The aesthetic argument belongs to this latter strand of natural theological thinking. It can start out with fairly modest assumptions about God's existence and his relation to the world, and, through the phenomenology of aesthetic experience, ratchet up to more knowledge about the creative intentions of God by taking the artistic design stance. Within a theistic worldview, our pervasive tendency to perceive beauty becomes intelligible as a result of God's design intentions, while our tendency to seek beauty can be explained as a quest for God. We have seen that it is possible for nontheists to enjoy these experiences without a realist commitment to God's existence, although it remains an open question to what extent religious fictionalism is a coherent, live option.

8   The Argument from Miracles and the Cognitive Science of Religious Testimony

Can one infer the truth of theism, or more specific religious doctrines, from the testimony to miracles? The argument from miracles, also known as the historical argument, answers in the affirmative. It has been severely criticized, notably by Hume (1748), who asserted that no one is ever justified in accepting the testimony to miracles. The philosophical study of the argument from miracles consists of three interrelated questions:

• What do we mean by "miracle"; is this a coherent, intelligible concept (the semantic question)?

• Under what conditions could one accept the testimony to miraculous events (the epistemic question)?

• Can the testimony to miracles be used to argue for particular religious beliefs (the argumentative question)?

Answering "yes" to these questions leads to claims of increasing strength. Clearly, the argumentative question cannot be answered positively if one is never entitled to accept the testimony to miraculous events, and accepting such testimony is only possible if miracles form a coherent, intelligible concept.

In this chapter, we examine these interrelated questions from a cognitive perspective. As we shall see, a better insight into the cognitive basis of our understanding of miracles and their testimony can throw new light on the apologetic function of miracles and on their use as premises in religious arguments. We rely on evidence from CSR and the cognitive science of testimony to uncover psychological mechanisms that underlie the testimony to miraculous events. We will argue that miracles can be best understood as events that violate intuitive ontological assumptions; such events are termed *minimally counterintuitive*. Combined with historical definitions of miracles, this conceptualization leads to a cognitive view of miracles that is coherent and intelligible. Combining empirical evidence and philosophical work on testimony, we explore whether it is ever reasonable, and if so, under what conditions, to believe in miracles from testimony. In the last section, we will examine what this signifies for the argumentative question.

The Argument from Miracles

The argument from miracles differs from other a posteriori natural theological arguments (e.g., design, cosmological, and moral arguments) in several respects. Most forms of the argument aim to establish not just theism but a more specific religious belief, in particular Christianity (see, e.g., McGrew and McGrew 2009). The Christian argument from miracles accepts the testimony to the bodily resurrection of Jesus of Nazareth (henceforth referred to as "the resurrection") and infers from this the truth of specific Christian doctrines, such as the Incarnation (the belief that Jesus is the second person of the Trinity). The argument from miracles is less dependent on philosophical conceptual analysis than some other natural theological arguments, as it critically relies on historical evidence. While one can formulate, for instance, the cosmological argument without knowing anything about contemporary cosmology, the argument from miracles depends on the interpretation of early Christian sources, which relies on textual, cultural, and historical analysis. Since the seventeenth century, the argument from miracles has usually been treated as part of a cumulative religious argument (Taliaferro 2005), as belief in miracles seems to require some prior belief in the existence of God. As John Stuart Mill (1889, 410) already pointed out, "If we do not already believe in supernatural agencies, no miracle can prove to us their existence. The miracle itself, considered merely as an extraordinary fact, may be satisfactorily certified by our senses or by testimony; but nothing can ever prove that it *is* a miracle: there is still another possible hypothesis, that of its being the result of some unknown natural cause."

The argument from miracles is currently not considered a strong natural theological argument. It has few contemporary defenders. One reason for this is that it is now mostly known in terms of its treatment in Hume's (1748) essay *Of Miracles*. In this essay, hailed as a showpiece of astute philosophical reasoning, Hume argued that it is never reasonable to rely on the testimony to miracles. Other nontheist philosophers of religion (e.g.,

Mackie 1982) agree that Hume has decisively buried all hopes for arguing from the testimony of miraculous events to the truth of particular religious viewpoints. Unfortunately, this attitude has isolated the argument from miracles from its original historical and dialectical context, giving the false impression that it was simultaneously brought up and refuted by Hume. To put it in its historical context, Hume's essay can be situated in an ongoing debate on miracles that had already reached its apex in the 1720s: in his historical treatment of discussions on miracles during the seventeenth and eighteenth centuries, Robert Burns (1981) positions *Of Miracles* at the tail end rather than the vanguard of this debate.

Historically, the testimony to miraculous events had an important apologetic function, not just in Christianity but in other religious traditions as well.[1] For members of the early Christian church, the argument from miracles was vital to support the reasonableness of their faith. Paul, for example, took the historicity of the physical resurrection of Jesus as absolutely crucial for establishing the reasonableness of the emerging Christian faith:

If there is no resurrection of the dead, then Christ has not been raised; and if Christ has not been raised, then our proclamation has been in vain and your faith has been in vain. ... If Christ has not been raised, your faith is futile and you are still in your sins. (1 Corinthians 15:13–14, 17)

In scholastic philosophy, miracles were considered as important *motives of credibility*, reasons that could be offered to unbelievers to defend the truth of Christianity. Miracles maintained this apologetic function well into the eighteenth century (Harrison 1995). It is in this context that Hume's ultimate aim in *Of Miracles* can be understood: he did not want to argue against the metaphysical possibility of miracles but, rather, against the reasonableness of belief in miracles from testimony. If such belief is unwarranted, the foundational and apologetic function of miracles crumbles. Even though the argument from miracles is at present not a popular natural theological argument, its historical prominence and apologetic role make it interesting from a cognitive point of view.

Defining Miracles from Historical and Cognitive Perspectives

Hume (1748, 181), in an unnumbered footnote, famously defined a miracle as "a transgression of a law of nature by a particular volition of the deity, or by the interposal of some invisible agent." Contemporary philosophers follow this definition as well; for example, Swinburne (1968b, 320) defines a miracle as "a violation of a law of Nature by a god." The apologetic function of miracles predates our scientific understanding of laws of nature, a view that only emerged within the context of seventeenth-century natural philosophy.

How then were miracles understood prior to that time? It is useful to consider the etymology of "miracle," which derives from the Latin verb *mirari*, meaning to marvel or to be amazed at (a particular object). Augustine ([5th c.] 1953, 320), the first to attempt a definition of miracle, took this subjective approach: he defined miracles as "something strange and difficult which exceeds the expectation and capacity of him who marvels at it." Miracles are not contrary to nature, but rather to our knowledge of its workings. In this way, Augustine located the miraculous in the experience of the individual who witnesses the miracle or learns of it through testimony. Aquinas ([13th c.] 1975, section 103) attempted to provide a definition of the miraculous as a property of an event rather than an effect upon the observer. Accordingly, a miracle is an action that goes beyond the inherent powers of created substances, such as human beings. Like Augustine, Aquinas did not think that miracles were contrary to nature. It is not against nature if God chooses to work in a way that is different from what nature is accustomed to; what he brings into being in it will become natural to it ([13th c.] 1975, section 100). One reason why ancient and medieval concepts of miracle were so different from the modern one is that the then-notions of "nature(s)" did not lend themselves to the idea that there could be a comprehensive system of explanation for events in the material world. This was because authors back then believed that matter is inherently chaotic (e.g., Platonism) or that each kind of thing has its own specific nature that requires a distinct form of explanation (a more Aristotelian view).[2]

The view that miracles violate the laws of nature only emerged during the early modern period in the writings of natural philosophers such as Robert Boyle and John Wilkins, who began to develop a mechanical view of the world as governed by orderly and lawlike processes. They regarded God as the orchestrator of the laws of nature, who was not bound by these laws but occasionally intervened to bend them. Thus, miracles became exceptional events within the uniformity of nature, brought about by God to point to the truth of religious beliefs. God's motivation for bringing about miracles was to inform those who did not engage in natural philosophical reasoning or theological study about particular religious truths. Unfortunately, this view, as it developed in the seventeenth century, led to a rather unstable concept of miracles, which was parasitic on the concept of immutable and stable laws of nature, while at the same time subverting the very concept of laws of nature by violating them (Pannenberg 2002). Miracles are simply an incoherent notion, and, for that reason, reports of miracles can always be ruled out a priori (Flew 1985). Hume, in keeping with his empiricist philosophy, regarded laws of nature as contingently true universal generalizations, based on induction (Taylor 2007). Since on this view miracles are contrary to inductive probability, their probability is always prima facie low; otherwise they would not be violating an established law of nature in the first place. This conceptualization of miracles as violations of laws of nature makes them either incoherent or vastly improbable.

As we have seen, from Augustine onward there has been an alternative way to define miracles in terms of their effects on the subjective experience of those who witness them or receive testimony to them. The natural philosopher Samuel Clarke (1728, 372–373), in contrast to his contemporaries, insisted that the key feature of miracles is their unusualness since God's actions are at work in all natural events, including the perpetuation of the laws of nature: "Either *nothing* is miraculous, namely if we have respect to the power of God; or, if we regard our own power and understanding, then almost *every thing*, as well what we call natural, as what we call supernatural, *is* in *this sense* really miraculous; and 'tis only *usualness* or *unusualness* that makes the distinction."

Paul Tillich (1953, 128–131) harkened back to the New Testamentic (in particular, Johannine) concept of σημεῖον (*semeion*, "sign"). He regarded miracles as sign-events, extraordinary occurrences with religious significance that have a "shaking and transforming power" over those who witness them, even if their cause has a perfectly natural explanation. This subjectivist concept of miracle does not suffer from the conceptual instability that the law-based account is vulnerable to. Therefore, some contemporary theologians (e.g., Pannenberg 2002) favor the subjective account and do not define miracles by reference to objective laws of nature. However, the subjective account, at least in its formulation by Augustine and Tillich, is somewhat too liberal—anything can be a miracle as long as it is subjectively experienced as a sign event.[3]

In what follows, we offer a subjectivist approach that avoids this problem by taking into account universal features of human cognition. According to CSR, our assessment of the unusualness of an event is not just a matter of individual taste but is constrained by universal, evolved features of human cognition. As we saw in chapter 2, developmental psychology indicates that humans are endowed with intuitive ontologies, by which our acquisition of knowledge and perception is structured. A provisional way to operationalize miracles in cognitive terms is to see them as exceptional, unexplained events that violate our intuitive ontological expectations and that have religious significance.[4] In this sense, miracles do not violate objective laws of nature but, rather, our subjective and intuitive understanding of them. Applying CSR to the resurrection, we can see why it is regarded as a miracle: it is both a violation of an intuitive ontology (intuitive biology) and an event with religious significance.

According to H. Clark Barrett, an intuitive understanding of death as the permanent and irreversible cessation of agency is part of our innate biological knowledge by virtue of its adaptive benefits. A live, sleeping lion is a potential danger, but a dead one is not. Conversely, a prey animal that is dead cannot run away; one that is merely asleep can still flee. To test this prediction, Barrett and Behne (2005) asked young children from Germany and from the Shuar a series of questions about an animal's capacities to act. In one condition, they told participants that the animal was asleep and in the other, that a hunter had killed it. The preschoolers were then asked whether the animal was still able to move or would still be able to hurt them. By age four, participants correctly understood that animals that are dead (but not those that are asleep) can no longer act. Bering and Bjorklund (2004) enacted a story where a mouse was eaten by an alligator. They asked preschoolers and slightly older children whether the mouse would ever be alive again. Even among the four-year-olds, 96 percent responded that the mouse would not be alive again. It would no longer need food and drink, and its senses would no longer work. Interestingly, in both studies, children thought that the dead agents still had psychological states. Even participants who explicitly endorsed that the mouse's brain was eaten and did not work anymore thought the mouse could still feel sad. This indicates that, at the level of our intuitive ontologies, psychological states are not regarded as biological properties. Hume (1748, 180) thought that events such as the resurrection are miraculous because they contradict our observations: "'Tis a miracle, that a dead man should come to life; because that has never been observ'd in any age or country." The developmental psychological evidence intimates that widespread observation in diverse places and times is neither necessary nor sufficient to establish what goes against perceived laws of nature. Given that many four-year-olds do not have repeated experience with dead agents, let alone have empirically observed what these can and cannot do, it seems plausible that an understanding of death as the permanent cessation of agency is not the result of induction, but a part of our intuitive biological knowledge. The resurrection is a violation of this, for instance, clearly indicated by Jesus eating a piece of broiled fish in the presence of the awe-struck apostles (Luke 24:41–43).

It may be more challenging to operationalize *religious significance.* One way to do this is to restrict miracles to those counterintuitive events that are explained as the result of divine agency. Experimental studies indeed indicate that people intuitively regard not just counterintuitiveness but also agency as an important element of religious concepts (e.g., Pyysiäinen, Lindeman, and Honkela 2003). While agency is not sufficient to categorize events as having religious significance or not, it is clearly an important element.

The Cultural Transmission of Minimally Counterintuitive Ideas

Boyer (1994, 2002) has developed an influential theoretical model to explain the transmission of religious beliefs. It starts out from the observation that most people do not invent their own religious beliefs, but usually acquire them from others. To account for the cross-cultural prevalence of some beliefs over others, we must therefore explain why it is that some concepts are transmitted and others are not. According to Boyer, memorability and ease of processing play an important role in cultural transmission. Cultural transmission is not a process by which information is faithfully copied from one brain to another. Rather, it is a reconstructive process. Acquiring concepts such as "socialism" through cultural transmission requires that a learner reconstruct it in her own mind. As a result, not everyone will have identical mental representations. Why then do culturally transmitted beliefs exhibit any stability at all? According to Sperber (1985), this is because learners do not start from scratch; they can build upon preexisting knowledge, including that supplied by intuitive ontologies. Based on her intuitive biological knowledge, a toddler who learns that a platypus is an animal can infer that it is self-propelled, needs food to sustain itself, and has offspring that resembles it. Intuitive ontologies constrain and guide the acquisition of concepts. We can therefore expect that concepts that have a poor fit with intuitive ontologies (e.g., scientific and mathematical concepts such as quantum mechanics and negative numbers) will be harder to reconstruct and thus harder to recall and transmit. On the other hand, concepts that are more in tune with our intuitive ontologies will be easier to reconstruct, remember, and transmit.

What makes supernatural agents and miracles so memorable? According to Boyer, such concepts strike a cognitive optimum because they are counterintuitive, but only minimally so. Boyer uses the term *counterintuitive* in a specific sense. It does not simply mean bizarre, but refers to a violation of intuitive ontological expectations. As we saw in chapter 2, intuitive ontologies furnish us with a set of intuitive expectations about how entities in the world behave. Religious concepts violate a few of our intuitive ontological expectations, which makes them interesting and attention-grabbing, but they conform to most other intuitive ontological expectations, which makes them easy to remember and transmit. This gives rise to a cognitive balance between ease and strikingness. Ghosts, for example, violate intuitive physics in their ability to walk through walls and to appear and disappear at will. But their psychology conforms to our basic ontological expectations about persons: they have beliefs, desires, and a distinct personality. Only the properties that violate our intuitive ontologies need to be transmitted when we acquire the concept "ghost" (their unusual physical properties); the elements that accord with intuitive ontologies are tacitly assumed (their normal psychology), as these are the default position. Consider the following two short stories:

• There was a wedding where the hosts ran out of wine. The guests drank water for the remainder of the party.

• There was a wedding where the hosts ran out of wine. One of the company turned water into wine. The guests drank this wine for the remainder of the party.

Even to those unfamiliar with John 2:1–11, the second story will be more memorable, in spite of it being longer than the first, because it has an unusual, attention-grabbing element: a violation of our intuitive essentialism about natural kinds, in this case, water and wine. For the rest, it adheres to our normal expectations—but for the bad planning, it is an ordinary wedding. Justin Barrett (2000) has termed stories and concepts that violate intuitive ontologies in this restricted sense *minimally counterintuitive* (MCI) to distinguish them from concepts that violate many intuitive ontological assumptions.

Boyer's model predicts that MCI concepts are easier to transmit than other concepts. This has been confirmed experimentally. Barrett and Nyhof (2001) asked participants to reproduce stories that contained intuitive, bizarre, or minimally counterintuitive elements. After a delay, their participants were much more accurate in their recall of the MCI elements, such as shoes sprouting roots, than they were in remembering bizarre items, such as blue horses. Boyer and Ramble (2001) obtained comparable results with a study that probed recall of similar stories in Gabon, Nepal, and France. These studies also indicate that MCI concepts are more robust in the long term: after three months, MCI elements were twice as likely to be recalled as bizarre items.

Minimal counterintuitiveness is a central property of religious concepts, but it is not specific to religion. Several studies indicate that the most successful (secular) fairy tales and folktales have MCI elements, such as a talking wolf or a frog turning into a man, to take but two well-known examples from Grimm's collection of fairy tales (e.g., Norenzayan et al. 2006; Barrett, Burdett, and Porter 2009).

Miracle stories typically relate MCI events: they violate only one or at most a few intuitive expectations against a background of normal events. The resurrection recounts that someone died (in line with intuitive biological expectations) but became alive again (a violation of intuitive biology). The birth of the Buddha was accompanied by both a bizarre event (he emerged from his mother's side) and an MCI event: the newborn took a few steps, each of which sprouted a lotus flower (a violation of intuitive biology). The minimal counterintuitiveness of miracles may explain why accounts of them are successfully transmitted, but it does not explain why people come to believe that they occur. At first blush, as they violate intuitive expectations, MCI events should—all things being equal—be less credible than events that conform to our expectations.

Neuroscientific studies have examined the cognitive processing of religious MCI concepts drawn from mythologies across the world (Fondevila et al. 2012). In these studies participants hear short stories while their N400 is measured using event-related potentials, a technique that measures changes in electrical activity in the brain. N400 is a neurological signal in the central and parietal areas of the brain, which appears about 400 milliseconds after the presentation of a stimulus. Stimuli that are difficult to process semantically tend to elicit larger N400s than stimuli that are easier to handle. In this way, N400 correlates with processing fluency. Events that violate expectations typically elicit larger N400s. These studies reveal that religious MCI events that were previously unknown to the participants elicit smaller N400s than other MCI events. For example, a story where the moon emerged from someone's mind (religious) elicited a smaller N400 than a story where a house emerged from someone's mind (nonreligious). Fondevila and Martín-Loeches (2013) hypothesize that this greater ease of processing may make religious ideas more acceptable than other types of MCI information. A broader narrative context renders MCI elements even easier to process. Nieuwland and Van Berkum (2006) found that MCI information is easier to digest when it is presented in a narrative framework. They measured N400s for stories with MCI elements, such as a yacht that went to see a psychiatrist because it had emotional problems. When they fitted this anecdote in a broader narrative, MCI events elicited no N400 effect, indicating that discourse can neutralize the unexpectedness of an MCI occurrence. Just like we are not too surprised to witness Tom molding Jerry into a nail and hammering him into the kitchen counter (this fits within the story of "The Missing Mouse"), miraculous events are typically situated in a rich narrative framework, making them less alien and unsettling.

To understand why miracles are accepted as true in particular religious communities demands more than merely paying attention to their *content*. Cultural transmission is also guided by *context*, the social factors that mediate the spread and acceptance of religious beliefs. To better analyze how cultural transmission works, Gervais et al. (2011) draw a distinction between content biases and context biases. *Content biases* are cognitive predispositions that direct our attention to the contents of particular representations, what they are about, and how they fit with our preexisting knowledge. *Context biases* draw our attention to the circumstances under which representations are transmitted, who one's informants are, and whether the source of the information is perceived as reliable. The transmission of religious beliefs is guided by both types of biases. For example, content biases can explain why people worshipped Zeus, a powerful agent with access to socially strategic information, but we need context biases to explain why Zeus is no longer an object of veneration, a few thousand Greek Neopagans notwithstanding (Gervais and Henrich 2010).

Since the argument from miracles depends not on direct observation but on the testimony of presumed reliable witnesses, we will now look at the role of content and context biases in testimony. This helps explain not only the cognitive appeal of miracles but also under what circumstances such testimony may come to be accepted.

Reliance on Testimony to Miracles

Humans acquire a vast amount of factual information through testimony, arguably more than they learn through experience. Without testimony, we would not know such things as the date of our birth, the name of the first president of the United States, the composition of the solar system, or the role of viruses in causing disease. The extensive reliance on testimony is remarkable given that one often cannot verify testimonial information oneself. Evolutionary studies of communication (e.g., Gintis, Smith, and Bowles 2001) indicate that relying on communication is risky because it can be advantageous to deceive others. In such cases, either signals become so costly that the benefits of cheating disappear (costly signaling), or learners become increasingly adept at detecting dishonest signalers. Most human communication through testimony is relatively cheap, which means that it is not particularly effortful for the testifier to convey the information. Since children and adults acquire an impressive amount of reliable information through testimony, and since testimony is cheap, we can expect that humans are skilled at evaluating information they receive from others. The cognitive processes involved are thus of potential relevance for epistemology. To examine the reasonableness of believing in miracles on the basis of reports, we will therefore look more closely at the cognitive science of testimony.

Philosophers (e.g., Audi 2013; Fricker 2006) regard testimony as a form of *telling*, that is, a social way of transmitting information. Storytelling, recounting something that has happened in one's personal experience, and conveying factual information are all forms of telling. What makes testimony distinct from storytelling is that it has an implicit or explicit assertion that the telling is true. The literary format and style of the Gospels is that of the ancient biography, a historiographic genre that was widely practiced in the ancient world (Keener 2011a). Thus, one can regard these accounts as a form of testimony. Testimony usually occurs in a *social* context—it has a testifier, who provides testimony, and an intended audience. The audience may be indeterminate, as in ostensibly solitary forms of testimony, such as writing a dairy, where the intended audience is one's future self, or in forms of testimony where the testifier does not know the audience, as in the crafting of a scholarly monograph. The audience can be heterogeneous, for example, a diary writer may have in mind future generations as well as her contemporaries; the audience of Mark may have been law-observing Jews, but his Gospel quickly became read by gentiles as well.

In epistemology, there are two broad positions that combine descriptive and normative claims of how humans (ought to) gauge testimony: reductionism and antireductionism. Both accept that testimony is an important source of information, but they differ in whether it constitutes a basic source of knowledge, such as perception or memory, that is, whether we can trust what others tell us in the same way that we trust our own senses or memories. *Reductionists* argue that this is not the case: testimony is not a fundamental source of warrant in the way that perception and personal memory are. Rather, our reliance on testimony is ultimately reducible to other sources of knowledge, such as our past experience of the reliability of the informant who does the telling. A person is entitled to accept testimony to *p only* if she has an appropriate degree of independent justification that the testifier is credible with respect to *p*. In other words, one has to have some positive evidence of one's interlocutor's knowledge or honesty in order to accept her testimony. Trusting testimony without such positive evidence is to believe "blindly, uncritically," tantamount to gullibility (Fricker 1994, 145). Hume was an early proponent of reductionism:

As the evidence, deriv'd from witnesses and human testimony, is founded on past experience, so it varies with the experience, and is regarded either as a *proof* or a *probability*, according as the conjunction betwixt any particular kind of report and any kind of objects has been found to be constant or variable. There are a number of circumstances to be taken into consideration in all judgements of this kind; and the ultimate standard, by which we determine all disputes, that may arise concerning them, is always deriv'd from experience and observation. ... We frequently hesitate concerning the reports of others. We balance the opposite circumstances, that cause any doubt or uncertainty; and when we discover a superiority on any side, we incline to it; but still with a diminution of assurance, in proportion to the force of its antagonist. (Hume 1748, 177–178)

The overall normative claim Hume made in *Of Miracles* is that credulous trust in testimony is not warranted; however, he also made a descriptive claim (as shown in the quote above) that humans in fact adopt a cautious, critical attitude when they receive second-hand information.

Anti-reductionists argue that we are entitled to accept testimony without any positive evidence. This happens frequently. For example, when asked for your date of birth, you will simply give that date as a trusted, accepted fact, without having to rely on inferences about the reliability of the sources through which you came to this belief. Testimony is a fundamental and irreducible source of knowledge, similar to memory and perception. We rely on memory, perception, and testimony to acquire knowledge; we do not usually attempt to justify our reliance on these sources. Even though they can sometimes lead to error (e.g., false memories, visual illusions, spurious testimony), they are generally reliable knowledge-acquisition processes, which is why we are entitled to rely on them without justification (Burge 1993). Thomas Reid was an early proponent of this view:

It is evident, that in the matter of testimony, the balance of human judgment is by nature inclined to the side of belief; and turns to that side of itself when there is nothing put in the opposite scale. ... In a word, if credulity were the effect of reasoning and experience, it must grow up and gather strength, in the same proportion as reason and experience do. But if it is the gift of nature, it will be strongest in childhood, and limited and restrained by experience; and the most superficial view of human life shows, that the last is really the case, and not the first. (Reid 1764, 478–479)

Like Hume, Reid mixed normative and descriptive claims about testimony. He offered the early development of trust in testimony as evidence for the view that it is "a gift of nature," or as cognitive scientists today would call it, innate, or maturationally natural. To Reid, this innate trustfulness is warranted by the fact that humans are also generally truthful. Underlying testimony are "two principles that tally with each other" (Reid 1764, 474), the principle of veracity and the principle of credulity. The former is "a propensity to speak truth, and to use the signs of language, so as to convey our real sentiments" (Reid 1764, 475), the latter "a disposition to confide in the veracity of others, and to believe what they tell us" (Reid 1764, 477). A number of contemporary philosophers (e.g., Audi 2013) endorse Reid's claim that we are prima facie entitled to accept testimony as long as there is nothing in the opposite scale, such as when the testimony is internally inconsistent or the testifier is a known confabulator. The fact that we learn to distrust incoherent testimony or dishonest testifiers is not incompatible with antireductionism. Just like a person can learn to distrust her perception in particular cases (e.g., when looking at a carnival mirror or leafing through a book of visual illusions), she can learn to distrust testimony under some circumstances.

Cognitive psychologists have found that children and adults behave in line with antireductionism but that they are nevertheless sensitive to defeaters to testimony. A large body of empirical evidence (see, e.g., Lindsay 2008 for review) indicates that adults typically do not tag a belief with a mental note concerning its origin. Of course, in many instances, people can make educated guesses about the likely source of a memory; for example, someone who has never been to New York (or Paris) can infer that her mental image of the Statue of Liberty must come from photos or movies. But in many cases, memories and testimonial information can easily become blurred, and testimony can infect one's personal memories. This is perhaps most dramatically illustrated by experiments that fabricate memories (see Loftus 2003 for an overview). Elizabeth Loftus gave participants descriptions of childhood experiences provided by family members and encouraged them to remember these events. One of these stories was actually a pseudo-event that had never taken place (e.g., getting lost in a shopping mall at age five and eventually being rescued by an elderly person). About 25 percent of participants claimed to "remember" these childhood experiences, often adding embellishing details to their accounts.

An intriguing observation in child psychology is that young children persistently forget the source of testimonial information. Taylor, Esbensen, and Bennett (1994) found that three- and four-year-olds who acquire information through perception or testimony come to believe they have always possessed this knowledge. The children in this study were taught a new fact (e.g., tigers have stripes for camouflage). Even immediately afterward, the preschoolers insisted that they had always known these facts. Similarly, when shown that a litmus paper dipped in acid turns pink, the young participants insisted that they knew this novel fact for a long time, even though a pretest established that they were unaware of this. Their source amnesia cannot be attributed to a poor memory in general: children of the same age who receive a new toy can accurately report that they had just received it and that it was not long in their possession. A more plausible explanation is that young children are psychologically disposed to acquire knowledge through testimony and perception: the information received in this way is basic, in the sense that it is unreflective and not based on other beliefs. This leads them to the impression that they have always known these facts. Also, perhaps more crucially, children do not make a distinction between knowledge acquired through testimony (e.g., the tiger fact) and knowledge acquired through direct experience (e.g., the litmus fact). If this interpretation is correct, it provides significant empirical support for a central antireductionist claim, namely that testimony, like memory and perception, is a fundamental source of knowledge that does not depend on other cognitive processes such as induction or inference.

Nevertheless, the reductionist claim that children and adults are sensitive to defeaters when they evaluate testimony also receives support from the cognitive science literature. Paul Harris and colleagues found that even young children rely on a variety of cues to modulate their trust in testimony. For example, people who have given false testimony in the past are less likely to be trusted. In one experiment, two-year-olds were confronted with adults who systematically misnamed known objects and people who always correctly named them. In a later session, the toddlers were less likely to trust the unreliable speakers when they named unfamiliar objects and more likely to trust the adults who previously named the objects correctly (Koenig and Harris 2007). Perceived expertise is also an important element in our reliance on testimony. People rely extensively on testimonial sources such as manuals or the Internet, or seek expert advice when they are faced with a specific problem. Lutz and Keil (2002) found that even five-year-olds already grasp this division of cognitive labor. They understand that a plumber is better informed than a physician with regard to questions concerning the central heating and that the latter is to be consulted about health-related problems.

Interestingly, Harris found that children treat testimony to scientific and religious beliefs in a similar way. Both are unobservable, and both are invoked in explanations. Harris et al. (2006) asked six-year-olds from the Boston area how confident they were about the existence of religious or quasi-religious endorsed beings like God, Santa Claus, and the tooth fairy (which they regrettably lumped together in the analysis) and in the existence of scientific unobservables, such as germs, vitamins, and oxygen. The children were significantly more confident about the existence of scientific unobservables than they were about supernatural beings. These findings were replicated with Spanish children who grew up in religious households and attended religious schools, in religious populations in the US, and with Maya children (where Santa was replaced by wood spirits, whose existence is widely endorsed). Even there, children expressed less confidence about the supernatural beings than about germs or oxygen. Harris and Corriveau (2014) theorize that children do not find religious testimony intrinsically more doubtful than scientific testimony. Rather, across both domains, the quality of testimony differs: people almost never express doubt about the existence of germs, but they do about God. Even expressions of faith can indirectly raise doubts. Someone who affirms "I believe and trust in God, even in the face of adversity" is also conveying that there are reasons to doubt God's existence in the face of adversity; someone who says "We Mormons believe that the Book of Mormon is God's revelation" is signaling that outgroup members may not hold this belief. This sensitivity to controversy or lack of consensus may explain why American adults exhibit lower belief in evolution than people from other developed nations—American media, churches, and other channels openly express doubt about the factuality of evolution. Lack of consensus is an important doubt-raiser for testimony, and it seems to sway people away from an antireductionist stance of trust.

An additional reason for the discrepancy between belief in germs and God might be that scientific entities are salient explananda: parents frequently appeal to these invisible entities in everyday explanations such as "Wash your hands before dinner to remove the germs that you picked up at school or on the bus," or "Let's open a window to let some oxygen into this room," whereas entities like Santa Claus and the tooth fairy only appear in temporally limited circumstances. In mainstream culture, germs and oxygen are more culturally salient and thus more likely to be endorsed. In young children, belief in miracles is likewise modulated by cultural salience. Corriveau, Chen, and Harris (in press) asked five- and six-year-olds whether they thought characters in a story were real or fictional. The stories came in three variants: only natural elements, an ordinarily impossible event caused by divine intervention, or the same event brought about by magic. For example, Joseph was introduced as a man who could tell the king about impending storms through observing the clouds (naturalistic version), through dreams sent by God (divine intervention version), or by his own magical powers (magical version). The children had to decide whether Joseph was a real or an imaginary character. In the religious condition, where divine intervention was explicitly mentioned, children from religious households were far more likely to believe Joseph really existed than participants from a secular milieu. However, children from both types of background tended to think Joseph was a fictional character when his powers were attributed to magic, which after all is not culturally salient in either group.

The current empirical evidence indicates that testimony is a fundamental source of knowledge, similar to memory and perception (in line with antireductionism), but that children and adults are sensitive to cues for the reliability of informants (in line with reductionism). For acceptance of testimony, both context and content are important. How can we explain this pattern? Compared to other animals, humans rely to a great extent on information they have not acquired for themselves, for example, in their use of technologies. This is evident not only in Western culture, with its highly developed technology and science, but also in cultures with sparser technologies, such as the Inuit, whose traditional kayaks, harpoons, parkas, and igloos make optimal use of a very limited range of materials, basically animal bone, fat, fur, snow, and occasional driftwood (Richerson and Boyd 2005). Michael Tomasello (1999) terms this the ratchet effect of human cultural transmission: because we readily rely on information obtained through testimony instead of reinventing everything on our own, human culture can be cumulative.

Studies that simulate cumulative culture in the laboratory indicate at least two key factors that make humans such good social learners. In such studies, young children typically outperform other primates, such as chimpanzees and capuchin monkeys. Three- and four-year-olds, unlike chimpanzees, verbally instruct their peers in difficult tasks, which increases their performance; that is, they demonstrate a willingness and ability to provide testimony (Dean et al. 2012). They also pay attention to the solutions and instructions of others to solve these tasks; in other words, they exhibit trust in the information others provide (Horner and Whiten 2005). These observations are in line with Reid's proposal that truthfulness and trustfulness play a crucial role in human knowledge acquisition and may have an evolutionary basis. If these evolutionary considerations are correct, they provide further support for a qualified form of antireductionism, where we are prima facie entitled to trust testimony in the absence of defeaters such as the past unreliability of the informant or a lack of cultural salience.

Let us briefly consider the implications of this empirical literature for the argument from miracles. The Reidian claim that we often rely on testimony without positive evidence for the speaker's reliability and are justified in doing so is plausible in the light of evolutionary considerations and the psychological dispositions that people exhibit. Nevertheless, children and adults are sensitive to both context and content that can potentially defeat justification. Thus we have come to a qualified antireductionism, where one can rely on testimony without positive evidence but not if it does not cohere well with other information one has, such as cultural salience. This means that in a nontheistic worldview, testimony to miracles does not seem to cohere well with background information, whereas in a theistic worldview, it can.

Implications for the Argument from Miracles

In the previous sections, we have examined miracles from a cognitive perspective and looked at the mechanisms involved in their transmission through testimony. We will now assess whether it is ever reasonable to accept the testimony to miracles, and if so, whether such accepted testimony can be used to argue in favor of particular religious beliefs.

From an internalist perspective, a person is entitled to accept testimony to miracles if they cohere with one's worldview.[5] Within a theistic worldview, it is possible that miracles occur (Ward 2002). This does not mean that miracles will occur—after all, God might also choose to act only in accordance with our views of the workings of nature—but it does make them more plausible. As Alan Hájek (2008, 103) puts it, Hume and others have assumed "that miracles are rendered so improbable because they are utterly disanalogous to anything that we have experienced. ... This will cut no ice with a person who has high prior probabilities for God's intervening in the world in various ways."

Internalism need not result in gullibility, in the sense that one accepts all testified miracles. For the theistic internalist, testimony to miracles is only acceptable if they fit with one's accepted beliefs about God and the world, if they seem to effect an important divine purpose, and if they cohere with one's beliefs about divine action (Ward 1985, 141).[6] Those who do not accept supernaturalism, by contrast, cannot place divine action within their worldview. In the latter worldview, reports of the miraculous can only be either false testimony or evidence that one's understanding of how nature works is incorrect. The internalist perspective explains why atheist authors such as Stephen Law (2011) tend to agree with Hume that the probability of correct testimony to miracles is vanishingly small, whereas theistic authors such as Keith Ward (2002) argue that if God exists, one can expect that some of this testimony will be veridical. But because of the interdependence of the acceptance of miracles and theism, internalism does not provide fruitful grounds for an argument from miracles. If theism is true, "it will be possible for miracles to occur and to be so reliably recorded that expectations are upset. The existence of such reliable records would indeed constitute a worthwhile argument for the possible action of a god, even though it may be far from decisive on its own. Whether reliable enough records actually exist is a different question; the point is that they certainly *could* exist" (Ward 1985, 138). This line of reasoning requires an antecedent belief in the existence of God (or gods), for it is only within a theistic worldview that miracles enjoy high enough antecedent probability to overcome skepticism about their reports.

Are some miracles so extraordinarily well attested that they overcome the initial skepticism of the nontheist? This seems unlikely. Miracle stories are typically only endorsed if they fit within and promote culturally accepted (religious) beliefs. Despite their good documentation and lack of a conclusive naturalistic explanation, few non-Hindus were swayed by the recent report of the milk-drinking statues of Gaṇeśa and other gods of the Hindu pantheon, possibly also because this miracle was not situated in a clear narrative context (there was no purported reason for why the miracle took place).[7] Craig Keener (2011b, 661), after reporting hundreds of healing miracles that have no conclusive naturalistic explanation, finds that physicians are unwilling to accept a supernatural explanation, even if this means admitting that their patients were misdiagnosed and, in some cases, wrongly treated for several decades. Intriguingly, most of these documented healing miracles come from non-Western cultures rather than from the West, where physical occurrences are exclusively explained in scientific terms and where there is the idea that such phenomena need to be explainable in scientific terms. In non-Western cultures there is more explanatory pluralism: religious explanations can exist next to scientific ones, for example, the belief in South Africa that AIDS can be caused both by witchcraft and by a virus (Legare et al. 2012). The cultural environment thus alters the receptivity to claims of miracles. Reports of miracles might sway the agnostic or weak believer but will fail to convince unbelievers or those adhering to other religions.

In contrast to internalists, externalists require a connection between one's beliefs and the external world for those beliefs to be justified. For example, belief in the resurrection would be justified on an externalist account if there is an unbroken testimonial chain that connects reliable eyewitnesses to later generations. Interestingly, the superior recall of MCI events in testimonial transmission indicates that if a miracle such as the resurrection occurred, it would likely have been transmitted faithfully, that is, the person at the end of the chain of transmission would likely receive the message sent at the beginning of the chain in a relatively unchanged form. In matters of religious testimony, assessing the quality of a testimonial chain, especially of its earliest links and its continuity, is therefore of crucial importance. Islam has the oral tradition of the *ḥadīth*, sayings attributed to Muhammad, who directly received them from God. Sayings are composed of two elements, the *isnād* and the *matn* (the actual saying itself). The *isnād* documents the testimonial chain through which the saying was recorded, going up to Muhammad's closest companions. Most of the criteria for the veridicality of a particular *ḥadīth* are externalist (e.g., is the testimonial chain unbroken, number of reporters, reliability of reporters), and only few are internalist (nature of the text itself, i.e., whether it agrees with the rest of the Islamic corpus). For religious scholars, the justification of belief in the resurrection from an externalist perspective likewise depends on the extent to which the earliest recorded testimony is appropriately linked to it.

A surprisingly large number of New Testament scholars (about 75 percent, according to Habermas 2005) believe that the tomb Jesus was buried in was found empty. While an empty tomb does not establish the resurrection with certainty, it is an important precondition. Moreover, there is the testimony of sightings of the risen Christ from a large number of witnesses, included in the Gospels and Paul's first letter to the Corinthians (15:4–8). However, the fact that most New Testament scholars are theists should caution us to accept their historical evidence at face value. James Crossley (2006, 25), a secular New Testament specialist, bemoans the overtly theistic conclusions of many of his colleagues: "Would another discipline in the humanities seriously consider as historically reliable something as spectacular as someone literally rising from the dead?" Indeed, as theologian and New Testament scholar Dale Allison (2005, 125) writes, "Ideology plays such an important role in the scholar's conclusions. One suspects that, typically, most such 'conclusions' are largely known in advance." In response to this, theistic New Testament scholars assert that "supernatural explanations, while not suitable in every case, should be welcome on the scholarly table along with other explanations often discussed" (Keener 2011b, 1). It seems difficult to assess whether the New Testament scholars' openness to supernatural explanations is warranted in a non-question-begging way. Allison (2005) summarizes that the data are ambiguous: they are not of the quality or strength that they allow for an inference to the best explanation to the resurrection. They do not falsify it conclusively either. Thus, the externalist is not in a position to use the argument from miracles but also not to refute it. Even if New Testament researchers could conclusively demonstrate that the disciples found the tomb empty and saw the risen Christ, the resurrection could still be a cultural reality—something that the disciples experienced as real within their cultural interpretative framework but not necessarily something that would be experienced as real by scientifically literate people today (Craffert 2009).

Next to internalism and externalism, a third way to assess epistemic justification is *internalist externalism*, developed by William Alston (1988): to be justified in believing that p holds that believing that p is based on an adequate ground. Examples of grounds of belief include perceptual experiences, the subject's other beliefs, and testimony. The ground for a belief that p is not composed of the subject's total stock of experiences and beliefs but only of those that are actually taken into account in the formation of the belief that p. In spelling out what makes a ground adequate, Alston appeals to the (good but not infallible) truth-conduciveness of a ground being used relative to a particular belief. As a result of differences in grounding, A can be justified in believing that p, whereas B may not be justified in holding the same belief. Take two people who come to believe that Max is a collie. However, crucially, A's belief is grounded in knowledge about collie features (e.g., shaggy tricolor coat, bushy tail, pointed snout), whereas B thinks that any largish dog is a collie. Thus, A has an adequate ground for her belief and is justified in the belief that Max is a collie, whereas B does not have an adequate ground and therefore is not justified (even though, as it turns out, Max is a collie). In contrast to internalism, in internalist externalism the subject need not be able to make this ground explicit: "In order for my belief that p, which is based on ground G, to be justified, it is quite sufficient, as well as necessary, that G be sufficiently indicative of the truth of p. It is in no way required that I know anything, or be justified in believing anything, about this relationship" (Alston 1988, 281).

Let us illustrate what this can say about the justification for belief in the resurrection. Zoë is a Christian with an amateur interest for New Testament scholarship. She believes in the resurrection based on the following ground: testimony she received as a child from her parents, in Sunday school, and during religious education, the coherence of this with her other (theistic) beliefs, and her perusing recent books by New Testament scholars that conclude that the tomb was empty and that the disciples believed that Jesus had risen from the dead. Within an internalist externalist account, it is sufficient for Zoë to be justified that this ground is indicative of truth. Is this the case?

As we have seen, in general, reliance on testimony leads to the acquisition of knowledge. We have provided several reasons for why this is the case. There is an evolutionary basis for the role of testimony in human knowledge acquisition that would be inexplicable if testimony were mostly unreliable. Next to this, people are able to monitor the quality of the testifier. Zoë has acquired a body of knowledge through the testimonial channels of parents and education, so it seems that prima facie she is justified in believing that the resurrection occurred based on this. It might well be that testimony in this case fails to meet this requirement, as when Zoë accepted false beliefs about Santa Claus from her parents. But given that we cannot assume a priori that this is the case here, Zoë's acceptance of the testimony seems prima facie justified. It might well be that the first links in the testimonial chain that led her to believe in the resurrection were not justified in their belief. For instance, the tomb may have been found empty for a purely naturalistic reason, leading to a mistaken belief among the disciples.[8] But what we are concerned with here is not whether the disciples' beliefs are justified but whether *Zoë's* beliefs are. An analogy from science might illustrate this: in 2010, it was revealed that Marc Hauser, a well-known Harvard comparative psychologist, fabricated data in experiments with monkeys (Carpenter 2012). Prior to this revelation, most members of the scientific community believed that the data he provided were genuine and, accordingly, accepted conclusions that followed from them, for example, that New World monkeys are capable of detecting grammatical structures. Even though Hauser was not justified in drawing these conclusions based on his fabricated data, other scientists, who learned about this through testimony, were justified in believing his conclusions because the transmission of scientific results through well-monitored channels of testimony (peer-reviewed papers) is generally truth-conducive; given scientific practice, the other scientists' beliefs were false but justified.[9]

Similarly, even if Zoë's beliefs are false (for instance, because the beliefs of the early Christians were a cultural rather than a scientific reality), they may still be justified, given the role of testimony in human knowledge acquisition. For the same reason, Zoë's acceptance of New Testament scholarship may be justified, since trust in expertise is, in general, truth-conducive, which is all that is required for internalist externalism. It might indeed well be, as Crossley (2006) states, that the judgment of theistic New Testament scholars is clouded to the extent that they draw unwarranted conclusions, but this is not at issue here. What we are concerned with is whether Zoë (not the New Testament scholars) is justified in believing that the resurrection took place based on her perusal of books on the New Testament. She lacks the expertise to independently assess this; for example, she does not read Koine Greek, the language in which the New Testament is written, and she has no hands-on experience in biblical archeology. However, does this mean that she has to accept these conclusions at face value? Not necessarily. Goldman (2001) has pointed out that while an amateur like Zoë might not have any reasons for believing that *p*, she might have *reasons for believing that the experts have good reasons for believing that p*. So, even if Zoë had no good reasons of her own to believe that the tomb was empty (and we know she has these reasons from the testimony from her parents and education), she can, by recognizing the expertise of New Testament scholars, believe that the experts have good reasons to believe that the tomb was empty and that the resurrection is a historical fact.

Note that Zoë's ground leads her to justifiably believe in the risen Christ but that other people with different grounds might reach divergent conclusions. A person who has been raised an atheist or within a different religion does not have access to Zoë's testimonial evidence; a biblical scholar who disagrees with the majority opinion within his field need not defer to his colleagues' views. To the internalist externalist, the miracle of the resurrection can be a compelling way to testify to the truth of Christianity, but given that the ground on which this argument is based will vary between subjects, this cannot be generalized to a compelling argument from miracles. So while we cannot in general answer the argumentative question in the affirmative (i.e., one cannot argue for the truth of particular religious beliefs from the testimony to miracles), we can conclude that people can be justified in believing in miracles.

Summary

The argument from miracles requires that three conditions be met. First, the concept miracle needs to be coherent and intelligible. Using an Augustinian, subjectivist approach updated with insights from CSR, we have provided a coherent miracle concept, according to which miracles are MCI events with religious significance. Second, we have explained why people find the testimony to miracles acceptable, at least under some conditions. Our review of the cognitive science of testimony indicates that people pay attention to the reliability of informants, and, to a lesser extent, to the content of testified information. The context in which miracle accounts are transmitted is more important than their contents. Third, the testimony to miracles cannot be used to argue for the truth of particular religious beliefs, except if one already accords a prior plausibility to these beliefs. From an internalist and an internalist externalist perspective, believing in miracles can be justified for theists.

However, the argument from miracles, in order to also be compelling to those who do not believe its conclusion antecedently, requires that acceptance of testimony to miraculous events is justified in an externalist way as well, for instance, by a reliable chain of transmission from the earliest Christians to today. We have seen that this is not the case, since it is impossible to assess the veridicality of the earliest eyewitness testimonies that led to the claim of the resurrection in the New Testament. Moreover, it is not clear whether the consensus among New Testament scholars concerning the empty tomb and sightings of the risen Christ should be taken as expert opinion or as the result of a preexisting theistic bias.

9   The Natural History of Religion and the Rationality of Religious Beliefs

Causal accounts of the origins of religious beliefs continue to generate questions about their justification. If a causal explanation targets beliefs in supernatural entities, does this damage the epistemic standing of these beliefs? Traditionally, the causal origins of religious beliefs and their justification have been treated as two separate questions. Mindful of the evidence surveyed in this book, we argue that this strict separation between causal origins and justification cannot be maintained. The causal origins of religious beliefs are also relevant for natural theology. This chapter considers evolutionary debunking arguments against religious belief and, more specifically, against natural theology. We argue that CSR cannot straightforwardly provide a debunking account of natural theology and religion because of the difficulty in choosing an appropriate level of explanation of how cognitive capacities generate religious beliefs.

Natural History of Religion and Justification

Whether or not CSR debunks religious beliefs, it is worth bearing in mind that it grew out of an intellectual tradition that sought to discredit religious belief. Anthropology and psychology of religion, two precursors of CSR, were the intellectual heirs to an atheist tradition of attempts to explain away religion, by explaining it either as a purely social fact (Durkheim 1915) or as a purely psychological phenomenon (Freud 1927). While precursors to anthropology of religion, such as Bernard Fontenelle (seventeenth century), ostensibly exposed only non-Western religious beliefs and practices as products of human primitive irrational thought, they indirectly also challenged Christianity by drawing comparisons between its beliefs and "primitive" myths and rituals. Precursors to psychology of religion, such as Anthony, Third Earl of Shaftesbury (eighteenth century), attributed religious belief ostensibly to superstition and, by implication, any form of theism to gullibility, fear, and even psychological abnormality (see Stark 1999 for a historical overview). The early disciplines of anthropology and psychology of religion had clear normative agendas as well; they not only wanted to describe the causal origins of religious beliefs but also, by exposing them, help to clear the way for a more enlightened, rational way of thinking. It is telling that Auguste Comte (1841) identified the most primitive state of thinking with *l'état théologique*, the theological or religious stage where human culture is in the grip of hallucinations and passions, a state that allegedly would be replaced by a rational and scientifically informed understanding of the world.

As a result of scientific research, many preconceived ideas in the anthropology and psychology of religion came under fire. Real ethnographic research indicated the sophistication and complexity of indigenous religious belief systems, in contrast to the second-hand reports on which early anthropologists relied. For instance, Bronislaw Malinowski noticed that Trobriand Islanders (Oceania) only relied on magic when the situation was unpredictable and dangerous:

While in the villages on the inner lagoon fishing is done in an easy and absolutely reliable manner by the method of poisoning, yielding abundant results without danger and uncertainty, there are on the shores of the open sea dangerous modes of fishing and also certain types in which the yield greatly varies according to whether shoals of fish appear beforehand or not. It is most significant that in the lagoon fishing, where man can rely completely on his knowledge and skill, magic does not exist, while in the open-sea fishing, full of danger and uncertainty, there is extensive magical ritual to secure safety and good results. (Malinowski [1925] 1992, 31)

In stark contrast to the a priori views on which Freud and other early psychologists of religion relied, experimental psychological research, for instance by Gordon Allport, did not claim that religious believers were deluded or psychologically abnormal. The scientific study of religion moved away from promoting a secular ideology to become a purely descriptive science.

Nevertheless, with the development of CSR from the 1980s onward, there is a renewed debate on whether explaining religion is tantamount to explaining it away. This is perhaps best illustrated by the titles of popularizing books such as *The God Delusion* (Dawkins 2006) and *Breaking the*

Spell (Dennett 2006). To look at the implications of causal accounts of religion for their justification, we will here consider two well-known examples, Hume's *The Natural History of Religion* and James's *The Varieties of Religious Experience*. While both predate CSR, their broad explanatory structure is similar to it, in that they look for naturalistic causes for beliefs and practices directed toward a proposed supernatural realm.

Hume (1757, 1) provided a historical and psychological account of religious beliefs. He proposed that the core of religious belief is belief in an "invisible, intelligent power." The proposed sources for this belief were anxiety and uncertainty about the future, ignorance, and wishful thinking—none of which are reliable, truth-conducive cognitive mechanisms. However, Hume stated that the rationality of religious beliefs can be maintained by appealing to natural theology, which he professed was immune to his debunking strategy: while the causal origins of theistic belief are unreliable, theism can be maintained by an appeal to the rationality of natural theological arguments. Such arguments allowed people who at first only held shallow religious beliefs based on unreliable cognitive mechanisms to found their theism on a firmer ground.

For men, being taught, by superstitious prejudices, to lay the stress on a wrong place; when that fails them, and they discover, by a little reflection, that the course of nature is regular and uniform, their whole faith totters, and falls to ruin. But being taught, by more reflection, that this very regularity and uniformity is the strongest proof of design and of a supreme intelligence, they return to that belief, which they had deserted; and they are now able to establish it on a firmer and more durable foundation. (Hume 1757, 44)

Hume scholars disagree on whether this appeal to the primacy of natural theology was strategic or sincere. Was it a means to evade the charge of atheism, then still punishable by law, or did Hume really believe that *true religion*, a form of philosophically informed, thin monotheism was acceptable, whereas *vulgar religion*, the religious belief of ordinary people, was not? Benjamin Cordry (2011) argues that Hume was an atheist but not dogmatically committed to materialism and open to the intellectual virtues of a thin, philosophically informed theism. Peter Kail (2007) conjectures that Hume thought religious beliefs were unjustified; while focusing his attack on the religious beliefs of ordinary people, his critique was implicitly meant to attack philosophical theism as well. By contrast, Lee Hardy (2012) asserts that Hume's attack on vulgar religion did not extend to true religion, which he thought was rationally defensible through the design argument. In any case, Hume thought a separation between the causal origins of religious beliefs and their justification was feasible.

William James (1902) pondered the philosophical implications of nascent neuropsychological and psychiatric explanations for religion. To James, mystical experience was a central causal factor in the emergence and spread of religious beliefs. Since medical evidence at the time suggested a connection between mystical experience and medical conditions such as temporal lobe epilepsy, he wondered whether this meant that religious beliefs are unjustified.

Medical materialism finishes up Saint Paul by calling his vision on the road to Damascus a discharging lesion of the occipital cortex, he being an epileptic. It snuffs out Saint Teresa as an hysteric, Saint Francis of Assisi as an hereditary degenerate. ... due to the perverted action of various glands which physiology will yet discover. And medical materialism then thinks that the spiritual authority of all such personages is successfully undermined. (James 1902, 13)

He resisted this conclusion because medical materialism, broadly applied, will lead us to cast doubt on a wide variety of beliefs, an unwelcome conclusion.

Scientific theories are organically conditioned just as much as religious emotions are; and if we only knew the facts intimately enough, we should doubtless see "the liver" determining the dicta of the sturdy atheist as decisively as it does those of the Methodist under conviction anxious about his soul. ... They are equally organically founded, be they of religious or of non-religious content. (James 1902, 14)

Moreover, even if this causal history casts doubt on the legitimacy of religious beliefs, the real test of their rationality lies in their theology. The soundness of a theological system, James thought, could be tested in a way similar to that of scientific theories, through experiment and logic—again an appeal to the primacy of natural theology.

This separation of the natural history of religion and its roots in reason seems to shield religious beliefs from the possible corrosive effects of their causal origins. If one acquires good evidence, in the form of a good argument for one's belief, then one's belief, however initially acquired, would be rational. Still, there are at least two reasons to abandon this dichotomy. First, as we have seen throughout this book, the intuitions that fuel natural theological argumentation are not immune to causal, naturalistic accounts, as these intuitions result from the workings of normal human cognitive propensities. If natural theological intuitions are rooted in human cognitive dispositions, such as intuitive ontologies, they cannot be used as an independent gauge for the soundness of religious beliefs. Second, as we will see in the next sections, some philosophical accounts use the origins of a belief to cast doubt on its justification.

Undercutting and Rebutting Defeaters

We have many beliefs that seem at first justified, even though we have no idea about their origin. Few people know how perception works and yet routinely form justified beliefs based on what they see, hear, touch, smell, or feel. Once we do become aware of the source of a belief, we can make a judgment concerning its epistemic credentials. For example, a belief formed through perception (e.g., there is a giant rabbit in the room) has weak epistemic credentials if we consider it was formed under the influence of hallucinogens. According to internalist theories of justification, adding this information about the source of our beliefs to our total stock of beliefs can have implications for the justification of individual beliefs.

In our everyday lives, much of the evidence for accepting a particular hypothesis is what philosophers term *defeasible*, which means that it is open to revision. For example, if a carton with six eggs contains five eggs that turn out to be fresh, it seems reasonable (although not certain, since we are using induction) that the sixth egg is also fresh. The defeasible reason for this inference is that all eggs were probably laid on the same date; the five other eggs were fresh, so the sixth one will be fine as well. However, if we learn that the sixth egg was introduced from another box, the reason for accepting the initial hypothesis that the sixth egg is fresh becomes less compelling. The five fresh eggs provide defeasible evidence for the hypothesis that the sixth egg is fresh, but the knowledge of its separate origin provides a defeater for this. John Pollock (1987) draws a distinction between two kinds of defeaters: rebutting and undercutting. A *rebutting* defeater gives us reason to think the conclusion must be false. By contrast, an *undercutting* defeater does not challenge the conclusion directly, but makes us doubt that the evidence supports the hypothesis. Whether or not the evidence is rebutting or undercutting depends on what kind of box the egg comes from. If the sixth egg comes from another box where we do not know how old it is, the evidence is undercutting. However, if the sixth egg comes from a box that is three months old, we have reason to believe the egg will be bad, so the evidence in that case constitutes a rebutting defeater.

We can apply this framework to the implications of CSR for the justification of religious beliefs, making a distinction between undercutting defeaters, which cancel out the evidential support for theism, and rebutting defeaters, which provide us with reasons to believe that theism is false. In the literature on CSR and theism, most philosophical work has focused on the former, as the bulk of naturalistic explanations do not straightforwardly cast doubt on the view that theism is true.

Take the common consent argument. The prevalence of belief in God was traditionally regarded as strong evidence for theism. If God exists, it seems plausible that he would want people to know about his existence. He could bring this into effect by revealing himself through the words of prophets and by implanting in our minds an innate propensity to believe in him.

There is within the human mind, and indeed by natural instinct, an awareness of divinity. ... God himself has implanted in all men a certain understanding of his divine majesty. ... Yet there is ... no nation so barbarous, no people so savage, that they have not a deep-seated conviction that there is a God. ... From this we conclude that it is not a doctrine that must be first learned in school, but one of which each of us is master from his mother's womb and which nature itself permits no one to forget. (Calvin [1559] 1960, 43–46)

Kelly (2011) provides a contemporary formulation of the common consent argument: granting that majority opinion does not prove that God exists, it nevertheless provides defeasible evidence for theism. We defer to majority opinion in many circumstances, and arguably, this is often justified. To take an analogous case that happened to one of the authors: while in graduate school, Johan was working in a museum as a guard. On September 11, 2001, he overheard museum visitors talking about planes that had flown into the World Trade Center. Some time later, other visitors mentioned other attacks as well. Although it sounded highly implausible at the time, he formed the defeasible belief that terrorist attacks had actually taken place. This belief was solely based on common consent since, being on duty as a guard, he had no access to news media to check the truth of these claims. As this belief turned out to be justified, it seems that common consent, in the absence of any other evidence, can provide strong defeasible evidence for a given hypothesis, even if that hypothesis has an initial low prima facie probability. CSR provides an undercutting defeater for common consent since it can explain belief in God through mechanisms that do not appeal to God's existence in their explanation. For example, as we saw in chapter 6, belief in punishing deities increases cooperation between members of a group. Members of groups who believe in punishing deities are thus more successful than members of groups who do not, regardless of whether or not punishing deities exist. If successful, such defeaters *undercut* the evidential support for theism that cross-cultural theistic beliefs provide.

Some CSR explanations have been marshaled as *rebutting* defeaters for theism. Bloom (2009), for example, has argued that the evolutionary origins of religious beliefs can challenge the rationality of holding on to such beliefs: religion is a byproduct of several cognitive predispositions, which together give rise to a plethora of mutually incompatible religious concepts. The incompatibility of multiple religious beliefs provides us with a reason to think that God does not exist; that is, it is a rebutting defeater for the hypothesis that God exists. If one's religious beliefs are caused by the same cognitive propensities as those of others, in combination with the cultural environment where one is raised, then the fact that one ends up with, for example, Christian beliefs, is a matter of luck. Bloom uses this line of reasoning to argue against Plantinga's notion of religious beliefs as properly basic. As a result, Bloom thinks that his argument not only takes away the reason to believe one's belief is properly basic, but that it also undermines it. Note that Bloom is appealing to something akin to the argument from inconsistent revelations, which long predates the development of CSR. Does CSR add anything substantial to this argument? Perhaps one could interpret Bloom's argument as follows: common consent was regarded as strong evidence for theism, but CSR explains it without having to invoke God, by appeal to naturalistic mechanisms such as enhancing cooperation, and moreover, the mechanisms identified by CSR lead to a proliferation of religious beliefs. Thus, Bloom is appealing to CSR as both undercutting and rebutting theism. It is not entirely clear how successful these defeaters are in unseating religious belief, especially in the stronger sense of rebutting defeaters. As a matter of fact, a number of theists accept the CSR results and propose that God works through these naturalistic mechanisms to instill belief. For example, Clark and Barrett (2010) argue that agency detection and mental state attribution spontaneously give rise to belief in God and that therefore CSR provides empirical support for the existence of a *sensus divinitatis*.

Generalized Evolutionary Debunking Strategies and CSR

Externalist theories of justification do not focus on consciously held beliefs and reasons of an individual, but on the relationship between her beliefs and the world. Someone's belief can be justified even if she has no inkling about how this belief was formed, as long as there is a proper causal connection between her beliefs and the world. If the belief-forming processes that CSR identifies track truth, then the outputs of such processes (in casu, religious beliefs) are justified. By contrast, if such cognitive processes do not track truth, the beliefs that result from them are unjustified.

Evolutionary debunking strategies apply this externalist maneuver to try to debunk religious beliefs. There are two types of evolutionary debunking strategies, generalized and specific. According to the *generalized* debunking argument, we can discount the truth of any proposition *p* whenever we can offer an evolutionary explanation of why we believe that *p* because natural selection is an off-track process, that is, it does not track truth (Kahane 2011). This argument takes the following form:

1. We believe that *p* because we have an intuition that *p*, and there is an evolutionary explanation of our intuition that *p*.

2. Evolution is not a truth-tracking process.

3. Therefore, we are not justified in believing that *p*.

As we saw in chapter 5, despite its initial appeal, this strategy is not widely adopted because it would force us to reject a wide variety of other beliefs as well, including commonsense and scientific beliefs. This is the *problem of collateral damage* (Barrett 2007): if evolved cognitive mechanisms are not truth-directed in general, we are left with a global skepticism about the reliability of our cognitive faculties. As a result, generalized evolutionary debunking arguments end up undermining their own rationality. The theory of natural selection, like other scientific theories, relies on evolved cognitive faculties, and questioning those would lead to a position of self-defeat. If natural selection does not select for truth-tracking cognitive faculties, what reason do we have to trust the theory? Plantinga's (1993) evolutionary argument against naturalism develops a similar argument:

since natural selection selects only for fitness, and metaphysical naturalism offers no reason to believe that our cognitive faculties would be guided
toward true beliefs, we have no reason to believe that the scientific reasoning and gathering of evidence that supports natural selection is correct.[1]
Naturalists (both theists and nontheists) have responded to Plantinga's
evolutionary argument in various ways (see the essays collected in Beilby
2002). Regardless of the outcome of this debate, the evolutionary argument
against naturalism makes clear that *if* the naturalist wants to debunk religion on evolutionary grounds, she cannot coherently opt for a generalized
debunking strategy (see De Cruz et al. 2011 for discussion).

When will cognitive capacities track truth? It is important to bear in
mind that cognition is costly; brains consume disproportionate amounts of
energy. There are severe costs involved in brain growth: animals with larger
brains, mainly birds and mammals, typically need longer periods of parental care than animals with smaller brains because the former need a longer
time to mature, time that cannot be invested in more offspring. Human
brain tissue is extremely costly, requiring up to 20 percent of the basal
metabolism in adults and up to 40 percent in young children (Aiello and
Wheeler 1995). Other costs of large brains for humans are the extremely
long period of infant dependency and an increased need for high-quality
food, such as meat, next to a disproportionate share of obstetric problems
due to the conflicting demands of large brain size and a narrow pelvis
for bipedal locomotion. As Evan Fales (1996, 440) puts it: "*Homo sapiens*
has, more than any other species, specialized in intelligence as a survival
strategy. … Our heavy investment in big brains and otherwise mediocre
bodies makes it all the more unlikely that resources would be wasted on
elaborate belief-forming and processing mechanisms that have no practical
utility." Combined with the observations that an organism's beliefs influence its behavior and that an organism's behavior influences its fitness, it is
extremely unlikely that the majority of our beliefs would be entirely insensitive to the state of the world. Since cognition influences how animals
interact with their environment, it is only under rare circumstances that
cognitive capacities will be insensitive to the state of the world.

Many cognitive capacities do not track truth infallibly or under all conditions, but they are still sensitive to features of the world. Rationality is
bounded because organisms have to make trade-offs between accuracy and
efficiency. It is pointless to carefully plan the most optimal escape route

when faced with a hungry predator. We will term this bounded sensitivity of cognition to states of the world *world-sensitive*. For example, deer flies (genus *Chrysops*), which feed on blood, conceptualize their prey as middle-sized, not-too-tall objects that are in motion. These can be humans or cows but also, for instance, trucks or boats. The deer fly's cognitive resources do not allow it to differentiate between a moving truck and ambling cattle (Sullivan 2009). While deer flies do not track ontological categories accurately as they cannot distinguish a driving truck from a running deer, their tiny brains are nevertheless tracking real features of the world, namely objects more or less the size of cattle that appear to be self-propelled. Their perception is thus world-sensitive. To take another example, color vision varies dramatically across species because of different adaptive pressures. Many mammals only have two color receptors, frugivorous diurnal monkeys and apes (including humans) typically have three, birds have four (extending into the ultraviolet range), and mantis shrimps (*Odontodactylus scyllarus*) have as many as twelve, seeing the world in an inconceivably rich and nuanced Technicolor. All these forms of color vision are world-sensitive, as they track the reflective properties of physical objects in the environment.

From these evolutionary considerations, we can expect that most evolved cognitive capacities are world-sensitive. Metaphysical naturalists cannot simply surmise that all religious beliefs are false and then proceed to observe that evolutionary processes give rise to false religious beliefs. Evolutionary debunkers will have to specify why natural selection would lead us to adopt false beliefs in the religious domain, whereas it allows us to adopt true beliefs under many other conditions.

Specific Evolutionary Debunking Arguments against Religion

Specific evolutionary debunking arguments are aimed at debunking a particular domain of beliefs. Even if it is implausible that cognitive capacities are not globally truth-tracking, one may make this case for specific domains, as some have done, for instance, for moral beliefs (see chapter 6). Beliefs in such domains are then explained as products of one or more fitness-enhancing traits that function not by tracking truth, but by tracking some other property. The above-average illusion is a good candidate for such a trait. This is a well-established tendency in neurotypical people to overestimate their own (and their children's) positive qualities and to underestimate their negative qualities compared to others: most people are better than average drivers and they are kinder and smarter than average. Having such beliefs increases self-confidence and mental health, and correlates positively with the time and energy parents are willing to invest in their progeny (Wenger and Fowers 2008). This propensity does not track truth since it is highly unlikely that most people are above average, but the illusion boosts confidence through self-deception; it is an adaptive misbelief (McKay and Dennett 2009).

Evolutionary debunking arguments against religion point to the evolutionary origins of religious beliefs in order to cast doubt on their validity. Well-known examples of such arguments are aimed at debunking theism: even if God does not exist, we would still believe in him as a result of naturalistic mechanisms outlined by CSR. Such arguments take the following general form:

4. Humans have religious beliefs because they have evolved propensities to form such beliefs.

5. There is no proper causal connection between these evolved propensities to form religious beliefs and their truth-value.

6. Therefore, religious beliefs are unjustified.

Debunkers have adopted two types of strategies to defend premise 5: supernatural explanations are redundant because religious beliefs are explained naturalistically, or belief-producing mechanisms that are responsible for theistic belief are unreliable.

Many secular authors implicitly follow the first strategy, which appeals to parsimony. Dennett, for instance, assumes that a naturalistic account of religion obviates the need for supernatural explanations. This does not amount to a conclusive proof against supernatural entities, since, as Dennett (2006, 25) writes, "Notice that it could be true that God exists, that God is indeed the intelligent, conscious, loving creator of us all, and yet *still* religion itself, as a complex set of phenomena, is a perfectly natural phenomenon." Moreover, a theist could argue that if God exists, belief in him is justified because there *is* a proper causal connection between God's existence and our belief in him, as God, who created us, is the first cause of our capacity to believe in him (Murray 2008). Bering (2011, 196) is skeptical about such a move: "If scientific parsimony prevails, and I think it should, such philosophical positioning becomes embarrassingly like grasping at straws." However, Bering's inference to atheism from parsimony is less straightforward than it appears. It is doubtful whether scientific parsimony is a useful guide for deciding which beliefs to adopt in real-life situations. We are not guided by parsimony when, for instance, we infer the existence of other minds from human behavior. Also, Swinburne (2004) holds that the simplest explanation for the existence of the universe is to postulate a simple (undivided) being, God, who freely chooses to create and to sustain it—quite a different appeal to parsimony. Thus, the first strategy does not conclusively establish premise 5.

The second strategy focuses on the belief-producing mechanisms that have as their outputs beliefs in supernatural entities. Many discussions on the justification of religious beliefs consider whether the cognitive mechanisms responsible for those beliefs are reliable.[2] *Reliabilism* is a philosophical tool that examines whether a given belief is justified. Reliabilists argue that a belief is justified if there is a reliable connection between that belief and the external world. This condition is met if the belief is produced by a belief-producing cognitive mechanism that is reliably aimed at furnishing us with true beliefs and that avoids false beliefs. Reliably does not mean infallibly but invokes a probabilistic notion: such beliefs are justified if they are mostly true. A standard formulation of this principle is the following:

Reliabilist theory of justification S's belief in *p* is justified if and only if it is caused (or causally sustained) by a reliable cognitive process, or a history of reliable processes. (Goldman 1994, 309)

As reliabilism is an externalist strategy, *S* need not be aware of the reliability of the cognitive processes that lead to her beliefs. Heather recognizes her colleague John walking on the other side of the street. Visual perception in humans usually produces true beliefs. Therefore, according to reliabilism, Heather is justified in believing that John walks on the other side of the street. There are many different sorts of events (so-called token events) that fall under the banner of visual perception. Seeing someone in the far distance, at nighttime, in broad daylight, close by, cycling past in one's peripheral vision, are all token events that fall under the same type, namely, recognizing someone by visual perception. Yet it seems that in some of these token events, the belief that John is on the other side of the street is justified (e.g., in broad daylight), whereas in others it is less clearly so (e.g., in peripheral vision). How can a reliabilist determine in a principled way which type of cognitive process a given token event belongs to?

It is hard to fix on the type of cognitive process in a principled way. We could characterize Heather's perception as falling under the broad type of visual perception, or visual recognition of people, or visual recognition of people during the daytime, and so on. This is the *generality problem* for reliabilism. If we describe a belief as the output of a process, we need to specify what type of cognitive process it is. This is not always an easy task, and it turns out that choosing the type can have implications for whether or not a belief is justified.[3] Alston (1995) attempts to solve this problem by an appeal to psychology: the relevant type for any cognitive process is the cognitive process, identified by cognitive science, that is responsible for the actual belief formation. If we grant that cognitive science can help us identify such processes, it is still often the case that a given instance of belief formation will be subserved by several relevant psychological mechanisms. We could further specify the relevant types of cognitive process as those that best explain Heather recognizing John. Conee and Feldman (1998) think that it is dubious that there is only one best explanation for a given belief's formation, and this is undoubtedly right: there might be a mix of processes that together converge on recognizing John. Heather might recognize John by his characteristic hopping gait, by the fact that he always wears a black suit with a colorful tie, and by means of her face recognition module. In this case, there are several type processes, and one has to assess the reliability of each of these. In this case, we know the processes are reliable because people habitually recognize and greet colleagues in the street. But the problem is different when we do not know whether or not the belief in question is true, such as when we probe the reliability of religious beliefs.

Many debunking arguments against religious beliefs hinge on the assumption that the cognitive mechanisms that produce them are unreliable. To determine whether they are indeed unreliable, we need to examine whether the processes involved in religious belief formation generally produce true beliefs and avoid false beliefs. To provide a point of focus, we will here examine agency detection as a cognitive process that produces religious beliefs (Guthrie 1993). According to Guthrie, our agency detection system sometimes leads us to perceive false positives, observing an agent where none is present. The evolutionary rationale for this is that a false positive is less costly than a false negative, as the latter can result in a failure to detect a dangerous predator, a prey animal, or a potential mate, and the former only results in a small waste of time and energy. Justin Barrett (2004) expects that agency detection is hypersensitive; he has coined the term HADD (hypersensitive agency detection device) for our capacity to detect agents.[4] On the basis of this, Wilkins and Griffiths formulate the following debunking argument:

The idea that religious belief is to a large extent the result of mental adaptations for agency detection has been endorsed by several leading evolutionary theorists of religion. … These mechanisms are "hyperactive," leading us to attribute natural events to a hidden agent or agents. So none of the contemporary evolutionary explanations of religious belief hypothesizes that those beliefs are produced by a mechanism that tracks truth. … If the "hyperactive agency detection device" theory is correct, then people believe in supernatural agents which do not exist for the same reason that birds sometimes mistake harmless birds passing overhead for raptors. These beliefs are Type I errors, and they are the price of avoiding more costly Type II errors. (Wilkins and Griffiths 2013, 142–143)

Is agency detection unreliable? As Peter Godfrey-Smith (1991) already remarked, it would be pointless to have an agency detection system if it did not, at least sometimes, produce correct beliefs. The agency detection capacity *is* truth-oriented in its function; it indeed successfully detects other humans and animals. However, if it also produces many false beliefs, the process is unreliable, and we are unjustified in trusting its outputs, including the religious beliefs it produces. What we need to examine to assess Wilkins and Griffith's evolutionary debunking argument is how reliable this process is. Godfrey-Smith (1991) suggests that the actual rates of false positives depend on the kinds of agents that are detected and the kind of agent the organism is. A prey animal that fails to detect a predator dies (false negative). This cost is so great that we can expect animals with many predators (e.g., rabbits) to be hair-trigger sensitive to signals of agency, even taking into account the costs of false negatives (e.g., fleeing for nothing). By contrast, a predator that fails to detect a prey animal misses out on a meal, which is a lesser cost, whereas the cost of chasing an imagined prey animal in vain is still substantial. Therefore, a predator will also be prone to false positives, but less so than a prey animal. There is no unified type of HADD, but different types of agency detection with differing false positive rates. From this, we can predict that context will influence the tendency of agency detection to be triggered: someone who walks at night in a dark alley and feels uneasy is more prone to a false positive than when she walks on a busy street in broad daylight. Thus, the inference that agency detection is hypersensitive cannot be made without taking into consideration the conditions under which it operates.

Experimental evidence on agency detection suggests that humans are highly proficient and accurate detectors of agency. They identify agency in movements of simple geometric figures in computer animations, such as chasing and teasing (e.g., Scholl and Tremoulet 2000). They can also accurately distinguish these from physical non-agential motions, such as bouncing and falling (Martin and Weisberg 2003). People are better at spotting animals than lifeless objects: when they are presented with quasi-identical pairs of scenes in which one item has changed, they are more likely to discern this change if it is an animal (e.g., a small pigeon) than if it is a building or a conspicuous red truck (New, Cosmides, and Tooby 2007). This is a remarkable finding, especially given that people have been trained for years to monitor vehicles for sudden life-or-death situations in traffic. This suggests that agency detection is truth-oriented, and indeed, one can expect nothing less. Wilkins and Griffiths (2013, 142) point out that "the detection of agency ... has been a critical task facing human beings throughout their evolution." We should not be surprised that it is highly accurate. Sometimes agency detection does go awry, as when we hear wooden planks creak in the night and form the erroneous belief that there is a burglar in the house. But many more times, we form the belief that there is an agent when there actually is an agent. When debunkers of religious belief appeal to HADD, they are already assuming that the agent that is being detected (e.g., God) is of the false-positive kind, the sound in the night, and this is what Wilkins and Griffiths do. If there is a God, then detecting him may not be a false positive. Note that false positives are still possible even if theism is true. For instance, someone can be mistaken in believing that she discerns God's actions, presence, or communication. This is an ongoing concern for experientially oriented religious believers: is the experience really of God? Charismatic Evangelical Christians (e.g., the Vineyard Movement) train themselves in dedicated ways, through prayer, regular study of the Bible, and group discussion to learn to distinguish God's actions from normal, natural events and from the actions of other supernatural agents (Luhrmann 2012).

Debunking arguments based on the unreliability of agency detection fail to sufficiently specify the type of cognitive process that produces religious beliefs. Is the type process agency detection in humans, broadly speaking?

Is it more akin to the dark alley situation, when someone is alone in the woods and her insecurity leads her to perceive sylvanian spirits? Unfortunately, there have been no empirical studies about the conditions under which agency detection could give rise to religious beliefs. Without these, it is not possible to fix the type of agency detection that produces religious beliefs, which means that evolutionary debunking arguments like those proposed by Wilkins and Griffiths are at present inconclusive. Even if one assumes from the outset that agency detection is an unreliable belief-forming process, the generality problem remains. As Barrett (2007, 68) has remarked, cognitive scientists propose a variety of cognitive heuristics and strategies that yield belief in supernatural agents: "As HADD and the other cognitive mechanisms that promote belief in god do not work alone to generate beliefs, their accuracy cannot be evaluated in isolation." Given that many other cognitive mechanisms involved in religious belief formation are regarded as reliable in other contexts (e.g., causal cognition), it is not clear why the mere fact that religious beliefs have evolutionary origins would give us a reason to assume that they are unjustified. We conclude that the second strategy, an appeal to the unreliability of belief-producing mechanisms that are responsible for theistic belief, does not provide a conclusive debunking strategy either.

Does CSR Debunk or Vindicate Natural Theology?

Does CSR evidence for the naturalness of intuitions in natural theology debunk or vindicate natural theological arguments? Evans (2010) considers the empirical evidence on which natural theologians rely as *natural signs* that point to God's reality. Moral principles, the sense of awe felt for the cosmic order, and the fine-tuning of the universe are natural signs in that they point to specific aspects of reality that indicate the existence of God. Such natural signs can be compelling to the believer, even if the arguments that rely on them are weak: the believer can acknowledge the signs that lie at the core of these arguments and recognize their force. Evans appeals to empirical evidence to support his case for the existence of God. But as we saw in chapters 4 through 8, the natural world and its features do not speak for themselves. Rather, our intuitive ontologies (and other cognitive dispositions) give rise to an interpretation of the world and its features. Evans acknowledges that intuitions like these lie at the basis of our perception of natural signs. To examine implications of CSR for natural theology, one has to take the intuitions that underlie natural theological arguments into account when considering the total evidence relevant for theism.

To our knowledge, no one has yet outlined a debunking argument against natural theology based on CSR. Stewart-Williams (2005, 801) comes close when he argues that "we should be extremely cautious about accepting that there must be a causal answer to the question of why there is something rather than nothing" (see chapter 5). It may well be that our causal intuitions applied to the universe are incorrect, since causal intuitions were shaped at the level of interactions with middle-sized objects. Other natural theological arguments could be similarly criticized. Hume (1779) objected that the design argument relies on an analogy between human designers and artifacts on the one hand and a supposed divine designer and natural objects on the other. As we saw in chapter 4, the intuitions that lead us to infer a designer of natural objects have been shaped by, among others, our evolutionary history as designers and users of artifacts. While the evolved design stance does not cast direct doubt on theism, it can provide an undercutting defeater for the original evidence, that is, our perception of design and our inference to a designer.

On the other end of the spectrum, one finds evolutionary vindicating arguments. Vindicating arguments can be seen as the opposite of debunking arguments: they examine the causal history of a particular belief in a way that strengthens that belief. They harken back to Reid (1764), who proposed that commonsense beliefs, such as the belief in other minds, the past, and the mind-independent reality of the world are crucial for our ability to act adaptively. Their adaptive value provides evidence for their validity. Relying on Calvin ([1559] 1960]), Clark and Barrett (2010) apply this Reidian line of reasoning to support theism based on CSR. Humans across cultures have a *sensus divinitatis*, a robust, innate propensity to believe in God, just like they have a robust, innate propensity to believe in the existence of the external world. Some religious beliefs are properly basic; they derive their warrant from the view that God designed the human mind in such a way that it spontaneously forms these beliefs under a wide range of circumstances (Plantinga 2000). Clark and Barrett (2011) contend that CSR provides evidence for the proper basicality of religious beliefs: they are innocent until proven otherwise, and one is justified in holding them in the absence of arguments. As we have argued elsewhere (De Cruz and De

Smedt 2013), this Reidian account is in line with Reformed epistemological interpretations of the *sensus divinitatis* (e.g., Wolterstorff 1983).

Barrett (2011) expands this Reidian framework to provide a justification for intuitions that underlie natural theological arguments, such as the cosmological and design arguments. Such intuitions are the result of common-sense beliefs, and therefore, we are prima facie justified in believing them. Commenting on a paper of ours (De Smedt and De Cruz 2011) that lies at the basis of chapter 5, he concludes:

> Interestingly, if one adopts a Reidian epistemology, the conclusions we can draw from De Smedt and De Cruz's analysis might be stronger still. If, as De Smedt and De Cruz argue, all of the relevant intuitions for cosmological arguments are automatic deliverances of our cognitive equipment, we should extend to them the benefit of the doubt. These intuitions have a high *prima facie* probability of being true. The burden of proof is on those wishing to reject the intuitions. Hence, critics of cosmological arguments must do more than assert that they see no reason why the universe needs a cause, or even offer an alternative causal explanation for the universe's origin. They need to offer arguments that sufficiently unseat the natural intuitions that it both needs a cause and that a god is a satisfying cause. (Barrett 2011, 158)

One could object to this that causal cognition does not always yield reliable outputs. However, this is not a decisive objection to this vindicating strategy. If causal cognition works reliably—an assumption that is not implausible, given the influence of cognition on behavior, its costs, and its world-sensitive nature—one can maintain that the causal intuitions that drive natural theology are justified. They are justified because they are produced by a propensity to detect causes that reliably generates correct beliefs. (Remember that reliable is a probabilistic notion: beliefs are justified if the mechanism that produces them is mostly on the mark.)

As a vindicating strategy, appealing to similarities between natural theological intuitions and common sense encounters the generality problem. For example, it is not at all clear on what level of generality we should evaluate causal cognition. Barrett (2011) looks at it from a developmental perspective, regarding causal cognition as the output of dedicated intuitive ontologies in young children: since our causal intuitions are automatic deliverances of our cognitive system, they have a high probability of being true (a reliabilist strategy). However, invoking God as the agent who caused the universe to come into existence is not appealing to an ordinary type of cause. Perhaps our relevant type of cognitive process should be the discernment of causes in domains that are out of the ordinary. In chapter 5, we saw that this yields very different intuitions: once we leave the level of ordinary, middle-sized objects, it is not at all clear that the deliverances of causal cognition generate reliable outputs:

It is probably reasonable to think that causal cognition is appropriate within the conditions and circumstances for which it evolved. In this range of conditions, it may be a close enough approximation to be biologically useful. However, outside this range, it may simply be inapplicable. Indeed, intuitive causal cognition appears not to work at the micro level described by quantum mechanics, where the idea that every event has a cause appears not to hold. In addition, it is entirely conceivable that causal cognition is not applicable to philosophical questions far removed from the evolutionary needs of our ancestors. Take, for example, the question of why there is something rather than nothing. Just as causal cognition is apparently not applicable at the quantum level, it may not be applicable to questions such as this. (Stewart-Williams 2005, 801)

Stewart-Williams is skeptical about the soundness of cosmological arguments: once we apply the causal principle to the existence of the universe, our intuitions become unreliable. Barrett and Stewart-Williams draw divergent conclusions because they look at causal reasoning at two different levels of generality: causal cognition is a reliable cognitive process (Barrett) but perhaps not once we leave the domain of middle-sized objects (Stewart-Williams). As we have seen in chapter 5, the danger of the latter position is that it also argues against the possibility of science, which would make Stewart-Williams's position incoherent since evolutionary psychology (his field of expertise and the discipline on which he relies to formulate this debunking argument) is the result of scientific inquiry and, hence, involves an abundant application of causal intuitions in a domain that is not ecologically relevant.

Divergent intuitions about the reliability of the testimony to miracles (chapter 8) are similarly due to differences in generality. When we consider testimony broadly, as an adaptive human strategy, there are good evolutionary reasons to suppose that this strategy is reliable. With this broad picture in mind, anti-reductionists (e.g., Reid 1764) hold that we are prima facie justified in believing the testimony of others. Alternatively, one can look at this problem in a more fine-grained way and consider miracle accounts as a subset of testimony, in particular, testimony to unusual events. In this case, it is not clear whether trust in testifiers is a strategy that reliably yields true beliefs. Hume (1748) examined testimony to miracles in this narrower sense. It is hard not to be influenced by one's prior beliefs when trying to settle on an appropriate level of generality, and in this type of argument, this depends on whether or not one believes God exists.

The generality problem for natural theologians boils down to the following: how can I know that my intuitions about causality, design, and other features of the world are reliable when I apply them in a religious context? How do I know what are the relevant types of cognitive processes that I need to consider? Perhaps the problem can be avoided by simply arguing that natural theological intuitions lie in the domain of common sense. Michael Bergmann (2008) argues that relying on perception, memory, and other normally functioning human cognitive faculties is justified because our belief that these faculties are reliable is a properly basic belief. In other words, we have non-propositional evidence (in the form of basic beliefs) that our cognitive faculties are reliable. Applying this framework to the intuitions that underlie natural theology, Evans (2010, 177) holds that they are a reliable source of true beliefs: "If theistic natural signs operate naturally to produce basic beliefs in God, then such beliefs should, like other basic beliefs that are produced by natural faculties, be trusted unless there are good reasons to think that these belief-forming processes are unreliable." Indeed, a theist who already believes that the intuitions underlying natural theological arguments point to God's existence can argue that these intuitions provide a prima facie justification. However, if one does not believe that God exists, it seems reasonable to think that this broad level is not the correct level to gauge the reliability of these intuitions but, rather, that they operate in what Bergmann (2006, 199) has termed a "questioned source context," a context in which the subject is uncertain about the source's trustworthiness. One's prior assumptions about the existence of God mediate to an important extent the perceived reliability of cognitive faculties that are involved in the formulation of natural theological arguments—this holds for both debunkers and vindicators.

We can now see why theists and nontheists end up with very different conclusions about what we can gather from the evolutionary origins of religious beliefs and, more specifically, about the intuitions that underlie natural theological arguments. Nontheists start from the assumption that the natural world is all there is and attempt to explain religious beliefs by appeal to everyday, natural cognitive processes. One of the challenges for the metaphysical naturalistic worldview is to explain why such beliefs are widespread if their referents (supernatural entities) do not exist. It also has to account for the enduring appeal of natural theological intuitions, such as the observation of design, or the apparent normative force of moral evaluative judgments. A promising explanatory strategy is to argue that such intuitions result from the normal workings of our cognitive faculties, which spontaneously lead us to recognize causes, design, and beauty. In this view, the inference to a supernatural reality is a mistake, a misapplication of intuitions that have evolved for a different context than the one they are applied in when we engage in natural theology. While these cognitive processes are reliable in general, they are unreliable when applied to metaphysical questions. By contrast, theists begin with the supposition that God is responsible for the design of reality, including human minds. From this perspective, it is reasonable to state that our natural cognitive processes are working properly when they generate the intuitions that underlie natural theological arguments. If there is a God, and God wants humans to be aware of him, it seems plausible that he would make himself known to everyone and not just to those who engage in formal theology. The intuitions that govern our daily interactions with the world are reliable. In this view, it is likely that these intuitions are also on the mark when they point us to God's existence. Taking into account their respective outlooks, it seems that both theists and nontheists reach reasonable conclusions and are justified in holding them.

Summary

This chapter has considered the implications of CSR for the justification of religious beliefs. Natural histories of religion grew out of an intellectual tradition that attempted to debunk religious beliefs by exposing their causal psychological and cultural origins. With the professionalization of the field, this normative dimension disappeared, and natural histories of religion became purely descriptive.

Nevertheless, the results of CSR continue to elicit questions about what the origins of religious beliefs can say about their justification. We first examined how awareness of the causal origins of religious beliefs can provide undercutting or rebutting defeaters, for instance, by undermining the evidential value of the ubiquity of theism. We then considered two debunking strategies against religion. Generalized debunking arguments lead to a position of self-defeat. Specific debunking arguments encounter the generality problem: one must first specify the types of cognitive processes responsible for religious beliefs before one can assess their reliability. At present, specifying the type of cognitive process turns out to be a difficult endeavor, as it is mediated by one's prior assumptions about the existence of God. Finally, we considered whether CSR can throw light on the justification of natural theological arguments. We looked at both an evolutionary debunking and an evolutionary vindicating strategy for the intuitions underlying these arguments. The generality problem plays here too: choosing which type of cognitive process is relevant for assessing these intuitions once again depends on one's prior view on the existence of God.

Notes

1 Natural Theology and Natural History

1. All Bible citations are from the New Revised Standard Version.

2. Epistemology is the philosophical study of knowledge and justified beliefs and is particularly concerned with what knowledge is, how we can acquire it, and under what conditions beliefs are justified.

3. Tickets for this debate, held in Oxford on February 23, 2012, were sold out within minutes after their release. See https://www.youtube.com/watch?v=HWN4cfh1Fac for the debate.

4. There is disagreement about how the term religion should be operationalized, or indeed whether this term should be avoided altogether (see, e.g., Saler 2008). We will not attempt to define religion here, as there seems to be a sufficient level of consensus about which beliefs and practices can be termed religious, and this is enough for our purposes. Few people, for instance, would regard communism and feminism as religions, but most would probably accept that Shintoism and voodoo are.

2 The Naturalness of Religious Beliefs

1. Quite a few cognitive scientists have put forward such views independently from each other, e.g., the cultural recycling hypothesis (Dehaene and Cohen 2007), the massive redeployment hypothesis (Anderson 2007), and the cultural domain of evolved cognitive modules (Sperber 1996).

2. Baillargeon, Spelke, and Wasserman (1985) introduced the violation of expectation procedure to test whether infants younger than nine months have object permanence (i.e., the idea that objects out of sight continue to exist). Piaget (1929) had proposed that infants under this age do not have object permanence because they fail to retrieve an attractive toy when it is hidden under a blanket. However, in Piaget's test, this failure may be due to a number of other factors rather than a lack

of object permanence, such as insufficient motor control. To minimize task demands, Baillargeon and colleagues devised displays that only required infants to look longer when object permanence was violated. Using this procedure, they found that five-month-olds indeed have object permanence.

3. Elsewhere we have explored how intuitive ontologies affect scientific understanding (De Cruz and De Smedt 2007, 2012).

4. Ontology is the philosophical study of what is, in other words, what kinds of entities there are in the world and how different categories of entities are related to each other. Human category-based inference mechanisms are ontologies in the sense that they provide a set of expectations of how specific kinds of objects behave.

5. Medieval impetus theory holds that objects can only be set in motion when acted upon by an external force. When this happens, the object acquires an impetus (a force transmitted through motion), which gradually dissipates until it comes to a standstill.

3 Intuitions about God's Knowledge: Anthropomorphism or Preparedness?

1. Euhemerus's work, dating to the fourth century BCE, is lost, but it is mentioned in Diodorus Siculus ([1st c. BCE] 1939, 333, 335, 337).

2. All Qur'an citations are from Abdel Haleem's 2005 translation.

3. The ḥadīth is an oral corpus of attested narratives of the sayings, actions, and tacit approvals and accounts of the prophet Muhammad. In the Islamic world, these play an important role in jurisprudence and theology (Abdel Haleem 2008).

4 Teleology, the Design Stance, and the Argument from Design

1. We spent several years (intermittently) trying to find the original mention of this powerful image. In the end, we stumbled upon a 1981 edition of *Nature* in which an anonymous author wrote a piece entitled "Hoyle on Evolution," which to our knowledge is the first mention of this metaphor.

2. The creationist roots of ID are extensively discussed by Forrest and Gross (2004), who argue that creationists restyled themselves as Intelligent Design proponents to circumvent restrictions on religious teaching in American school curricula. After the landmark ruling of *Edwards v. Aguillard* in 1987, which had outlawed teaching creationism in American public schools, explicit references to God were cut out of creationist handbooks and replaced with references to ID, often crudely through simple cut-and-paste.

5   The Cosmological Argument and Intuitions about Causality and Agency

1. In the discussion of the causal principle, this chapter will leave aside inductive versions of the cosmological argument (such as Swinburne 2004), which are based on considerations of simplicity.

2. Hume believed that our causal intuitions are purely based on the perceived constant conjunction of events. As will be argued in more detail further on, contemporary cognitive science indicates that people are also able to infer causal relationships for unique events if they can identify a plausible mechanism that connects them.

3. This distinction is not exhaustive. As we saw in the previous chapter, children also have teleological explanations for events, where these are explained in terms of purposes. Whereas younger children do not explicitly account for those purposes in terms of a designer (i.e., an agent), older ones do. By middle childhood, events with perceived teleology are causally attributed to agents (Rottman and Kelemen 2012).

4. The survey was conducted in late 2011. Respondents were recruited through philosophy mailing lists and blogs. 75.8 percent of subjects were men, 24.2 percent were women. The religious self-identification of the respondents was 40.5 percent theists, 40.4 percent atheists, and 19.1 percent agnostic. While this distribution may not be representative for philosophy, it does allow for robust comparisons between theists, atheists, and agnostics. Some results of the survey have been published in De Cruz (in press).

5. The statistics were calculated using a non-parametric test for independent variables, the Kruskal-Wallis ANOVA. The result for arguments in favor of theism was $(df(2) = 397.2, p = 0.0001)$, the result for the arguments against theism was $(df(2) = 217, p = 0.0001)$, which are strong effects.

6. Of 802 respondents, 788 rated the cosmological argument; the remaining 14 participants indicated they were not sufficiently familiar with the argument. In this sample, theists were 23.12 times as likely as atheists and 6.93 times as likely as agnostics to rate the cosmological argument more favorably $(p < 0.0001$ for both statements).

7. In this sample, philosophers of religion were 1.53 times as likely as those who do not have this specialization to rate the cosmological argument more favorably $(p = 0.01)$. These ratios were computed using a logit model. Thanks to Robert O'Brien for help with this calculation.

6   The Moral Argument in the Light of Evolutionary Ethics

1. There is more to this than meets the eye. While calling is an altruistic (i.e., costly) act, which increases inclusive fitness by warning both offspring and other close kin

of an impending attack, it may also be a form of costly signaling, where the caller indicates to potential predators that it has seen them, thus discouraging attack (Blumstein 2007).

2. Divine command theories regard God (or the gods) as a source of moral obligation and prohibition, instructing people about what is morally permissible and forbidden (see Quinn 2000 for discussion).

3. We thank Olli-Pekka Vainio for this observation.

7 The Argument from Beauty and the Evolutionary Basis of Aesthetic Experience

1. The argument from religious experience proposes that we can derive a significant degree of epistemic justification for the existence of God from religious experiences.

2. This means that either the paintings are ancient or that they are part of an uninterrupted visual tradition that goes back to at least 40,000 years ago, since it would otherwise be very hard to explain how the painters could have made these pictures so accurately that paleontologists can describe the species they represent.

3. Nevertheless, some cultures are more egalitarian, and some extremely so, including hunter-gatherer societies such as the !Kung (Kalahari desert, southern Africa). Members of these cultures exert constant vigilance to prevent dominance hierarchies from forming, and successful hunters habitually belittle their own exploits (Cashdan 1980).

4. According to Plantinga (1993), "proper functioning" means that a capacity is working in line with its intended design in congenial conditions.

8 The Argument from Miracles and the Cognitive Science of Religious Testimony

1. Our treatment of the Christian argument from miracles is for the sake of providing a point of focus.

2. We thank Steven Horst for this observation.

3. To take an example from Roy Holland (1965): a child playing on the tracks is about to be run over by an approaching train, but the train stops right before it would have killed the child. It turns out that the driver had fainted due to a blood clot reaching his brain, and as he fainted, his hand hit the brake. While this event has a perfectly natural explanation, the child's mother continues to regard it as a miracle, even after she has discovered why the train has stopped. What

makes this a miracle is that she attributes a particular meaning to the event, a sign from God.

4. There are two other cognitive accounts that resemble ours and that we discovered relatively late while writing this chapter. According to Pyysiäinen (2002), miracles in a weak sense are all events that run counter to our intuitive ontological expectations. Miracles in a strong sense are counterintuitive events that are explained by counterintuitive agents and forces from particular religious traditions. Our cognitive account of miracles resembles this latter concept. Robert Shanafelt (2004) describes counterintuitive events that are ascribed to a supernatural cause as "marvels" and restricts the term "miracles" to events with a positive result caused by divine intervention.

5. See chapter 5 for a discussion of internalism and externalism. There are other factors next to coherence that can provide internalist justification; here we focus on coherence as one example.

6. For instance, if one adheres to a Protestant denomination, one need not accept miracles that came about through the intercession of saints, since most Protestant denominations reject saintly intercessions.

7. Sivaya Subramuniyaswami (2000, xi–xxxi) relates the event, which took place in Hindu temples across the world in late September 1995, and collected dozens of newspaper cuttings reporting it.

8. Historically, reasons that were offered include that they went to the wrong tomb or that the body had been taken away.

9. A minority of scholars (e.g., Sutton 2005; Littlejohn 2012) argue that only true beliefs can be justified. Most epistemologists, however, take these notions apart and argue that one can have a justified false belief.

9 The Natural History of Religion and the Rationality of Religious Beliefs

1. Plantinga does not formulate this argument because he believes evolution is an adequate explanation, but to argue that only a supernaturalist ontology can provide an externalist justification for our beliefs.

2. Some authors use reliabilism to examine which beliefs constitute knowledge or which beliefs have warrant (e.g., Plantinga 1993). We will look at the justification of religious beliefs here, but much of our discussion could also be applied to their status as knowledge or their warrant.

3. Conee and Feldman (1998) argue that it is impossible to fix the type process and that this is a fatal problem for reliabilism. While most epistemologists think this is not the case, it nevertheless remains difficult to determine the type process for

particular beliefs without already assuming the conclusion one wants to draw (i.e., that the process is reliable or not).

4. Guthrie thinks the agency detection system is especially attuned to perceiving other humans, whereas Barrett sees the capacity as more generally aimed at detecting any type of self-propelled being. Guthrie (personal communication) does not agree with Barrett's characterization of agency detection as hypersensitive. Rather, he thinks the system is finely tuned and (on balance, given the costs of false negatives, leading to some false positives) "gets it just right."

References

Abdel Haleem, M. A. S., trans. 2005. *The Qu'ran*. Oxford: Oxford University Press.

Abdel Haleem, M. A. S. 2008. Qur'an and hadith. In *The Cambridge Companion to Classical Islamic Theology*, ed. Tim Winter, 19–32. Cambridge: Cambridge University Press.

Adams, R. 1987. *The Virtue of Faith*. New York: Oxford University Press.

Ahn, W., C. Kalish, S. A. Gelman, D. L. Medin, C. Luhmann, S. Atran, J. D. Coley, and P. Shafto. 2001. Why essences are essential in the psychology of concepts. *Cognition* 82:59–69.

Ahn, W., C. Kalish, D. L. Medin, and S. A. Gelman. 1995. The role of covariation versus mechanism information in causal attribution. *Cognition* 54:299–352.

Aiello, L. C., and P. Wheeler. 1995. The expensive-tissue hypothesis. The brain and the digestive system in human and primate evolution. *Current Anthropology* 36: 199–221.

Aiken, N. E. 1998. *The Biological Origins of Art*. Westport, CT: Praeger.

Akerman, K., and T. Willing. 2009. An ancient rock painting of a marsupial lion, *Thylacoleo carnifex*, from the Kimberley, Western Australia. *Antiquity* 83 (March). http://antiquity.ac.uk/projgall/akerman319/.

Alexander, J. 2012. *Experimental Philosophy: An Introduction*. Cambridge: Polity Press.

Allen, C., and M. Bekoff. 1995. Biological function, adaptation, and natural design. *Philosophy of Science* 62:609–622.

Allison, D. C. 2005. Explaining the resurrection: Conflicting convictions. *Journal for the Study of the Historical Jesus* 3:117–133.

Allison, P. D. 1992. The cultural evolution of beneficent norms. *Social Forces* 71:279–301.

Alston, W. P. 1988. An internalist externalism. *Synthese* 74:265–283.

Alston, W. P. 1991. *Perceiving God: The Epistemology of Religious Experience.* Ithaca, NY: Cornell University Press.

Alston, W. P. 1995. How to think about reliability. *Philosophical Topics* 23:1–29.

Ambrose, S. H. 1998. Chronology of the Later Stone Age and food production in East Africa. *Journal of Archaeological Science* 25:377–392.

Anderson, M. L. 2007. The massive redeployment hypothesis and the functional topography of the brain. *Philosophical Psychology* 20:143–174.

Anderson, R. L. 1989. *Art in Small-Scale Societies.* Englewood Cliffs, NJ: Prentice Hall.

Anselm. (11th c.) 2000. Proslogion. In *Complete Philosophical and Theological Treatises of Anselm of Canterbury,* trans. J. Hopkins and H. Richardson, 88–112. Minneapolis, MN: Arthur J. Banning Press.

Apicella, C. L., F. W. Marlowe, J. H. Fowler, and N. A. Christakis. 2012. Social networks and cooperation in hunter-gatherers. *Nature* 481:497–501.

Apperly, I. 2011. *Mindreaders: The Cognitive Basis of "Theory of Mind."* Hove: Psychology Press.

Apperly, I., E. Back, D. Samson, and L. France. 2008. The cost of thinking about false beliefs: Evidence from adult's performance on a non-inferential theory of mind task. *Cognition* 106:1093–1108.

Aquinas, T. (13th c.) 1975. *Summa contra gentiles: Providence, Part 2* (vol. 3), trans. V. J. Bourke. Notre Dame, IN: University of Notre Dame Press.

Aquinas, T. (13th c.) 2006. *Summa theologiae: Existence and Nature of God* (vol. 2), trans. T. McDermott. Cambridge: Cambridge University Press.

Aquinas, T. (13th c.) 2008. Treatise on law, questions 90–91, 94–96, *Summa theologica.* In *Readings in the Philosophy of Law,* 2nd ed., ed. K. C. Culver, 31–46. Toronto: Broadview Press.

Aslin, R. N., J. R. Saffran, and E. L. Newport. 1998. Computation of conditional probability statistics by 8-month-old infants. *Psychological Science* 9:321–324.

Astuti, R., and P. L. Harris. 2008. Understanding mortality and the life of the ancestors in rural Madagascar. *Cognitive Science* 32:713–740.

Atran, S., D. L. Medin, and N. Ross. 2004. Evolution and devolution of knowledge: A tale of two biologies. *Journal of the Royal Anthropological Institute* 10:395–420.

Audi, R. 2011. *Rationality and Religious Commitment.* Oxford: Oxford University Press.

Audi, R. 2013. Testimony as a social foundation of knowledge. *Philosophy and Phenomenological Research* 87:507–531.

Augustine. (397–398) 1961. *Confessions*. Trans. R. S. Pine-Coffin. London: Penguin.

Augustine. (416) 2002. The literal meaning of Genesis. In *On Genesis*, ed. J. E. Rotelle, trans. E. Hill and M. O'Connell, 155–506. New York: New City Press.

Augustine. (5th c.) 1953. The usefulness of belief (*De utilitate credendi*). In *Augustine: Earlier Writings*, ed. J. H. S. Burleigh, 284–323. Philadelphia: Westminster Press.

Axelrod, R., and W. D. Hamilton. 1981. The evolution of cooperation. *Science* 211:1390–1396.

Ayala, F. 1970. Teleological explanations in evolutionary biology. *Philosophy of Science* 37:1–15.

Ayala, F. 2010. The difference of being human: Morality. *Proceedings of the National Academy of Sciences of the United States of America* 107:9015–9022.

Azari, N. P., J. Nickel, G. Wunderlich, M. Niedeggen, H. Hefter, L. Tellmann, H. Herzog, P. Stoerig, D. Birnbacher, and R. J. Seitz. 2001. Neural correlates of religious experience. *European Journal of Neuroscience* 13:1649–1652.

Baillargeon, R., E. S. Spelke, and S. Wasserman. 1985. Object permanence in five-month-old infants. *Cognition* 20:191–208.

Barrett, H. C. 2004. Design versus descent in Shuar children's reasoning about animals. *Journal of Cognition and Culture* 4:25–50.

Barrett, H. C., and T. Behne. 2005. Children's understanding of death as the cessation of agency: A test using sleep versus death. *Cognition* 96:93–108.

Barrett, J. L. 1998. Cognitive constraints on Hindu concepts of the divine. *Journal for the Scientific Study of Religion* 37:608–619.

Barrett, J. L. 2000. Exploring the natural foundations of religion. *Trends in Cognitive Sciences* 4:29–34.

Barrett, J. L. 2001. How ordinary cognition informs petitionary prayer. *Journal of Cognition and Culture* 1:259–269.

Barrett, J. L. 2004. *Why Would Anyone Believe in God?* Lanham, MD: AltaMira Press.

Barrett, J. L. 2007. Is the spell really broken? Bio-psychological explanations of religion and theistic belief. *Theology and Science* 5:57–72.

Barrett, J. L. 2010. The relative unnaturalness of atheism: On why Geertz and Markússon are both right and wrong. *Religion* 40:169–172.

Barrett, J. L. 2011. *Cognitive Science, Religion, and Theology: From Human Minds to Divine Minds*. West Conshohocken, PA: Templeton Press.

Barrett, J. L., E. R. Burdett, and T. J. Porter. 2009. Counterintuitiveness in folktales: Finding the cognitive optimum. *Journal of Cognition and Culture* 9:271–287.

Barrett, J. L., and F. C. Keil. 1996. Conceptualizing a nonnatural entity: Anthropomorphism in God concepts. *Cognitive Psychology* 31:219–247.

Barrett, J. L., and M. A. Nyhof. 2001. Spreading non-natural concepts: The role of intuitive conceptual structures in memory and transmission of cultural materials. *Journal of Cognition and Culture* 1:69–100.

Barrett, J. L., and R. A. Richert. 2003. Anthropomorphism or preparedness? Exploring children's God concepts. *Review of Religious Research* 44:300–312.

Barrett, J. L., R. A. Richert, and A. Driesenga. 2001. God's beliefs versus mother's: The development of nonhuman agent concepts. *Child Development* 72:50–65.

Barry, A. M. 2006. Perceptual aesthetics: Transcendent emotion, neurological image. *Visual Communication Quarterly* 13:134–151.

Bartal, I. B., J. Decety, and P. Mason. 2011. Empathy and pro-social behavior in rats. *Science* 334:1427–1430.

Basalla, G. 1988. *The Evolution of Technology.* Cambridge: Cambridge University Press.

Bateson, M., D. Nettle, and G. Roberts. 2006. Cues of being watched enhance cooperation in a real-world setting. *Biology Letters* 2:412–414.

Beauregard, M., and V. Paquette. 2006. Neural correlates of a mystical experience in Carmelite nuns. *Neuroscience Letters* 405:186–190.

Behe, M. J. 1996. *Darwin's Black Box: The Biochemical Challenge to Evolution.* New York: Free Press.

Beilby, J., ed. 2002. *Naturalism Defeated? Essays on Plantinga's Evolutionary Argument against Naturalism.* Ithaca, NY: Cornell University Press.

Bergmann, M. 2006. Epistemic circularity and common sense: A reply to Reed. *Philosophy and Phenomenological Research* 73:198–207.

Bergmann, M. 2008. Reidian externalism. In *New Waves in Epistemology,* ed. V. F. Hendricks and D. Pritchard, 52–74. Basingstoke: Palgrave Macmillan.

Bering, J. M. 2006. The folk psychology of souls. *Behavioral and Brain Sciences* 29:453–462.

Bering, J. M. 2011. *The God Instinct: The Psychology of Souls, Destiny, and the Meaning of Life.* London: Nicholas Brealy.

Bering, J. M., and D. F. Bjorklund. 2004. The natural emergence of reasoning about the afterlife as a developmental regularity. *Developmental Psychology* 40:217–233.

Bering, J. M., K. McLeod, and T. Shackelford. 2005. Reasoning about dead agents reveals possible adaptive trends. *Human Nature* 16:360–381.

Blackburn, S. 2001. Normativity à la mode. *Journal of Ethics* 5:139–153.

Blancke, S., and J. De Smedt. 2013. Evolved to be irrational? Evolutionary and cognitive foundations of pseudosciences. In *Philosophy of Pseudoscience: Reconsidering the Demarcation Problem*, ed. M. Pigliucci and M. Boudry, 361–379. Chicago: University of Chicago Press.

Blood, A. J., and R. J. Zatorre. 2001. Intensely pleasurable responses to music correlate with activity in brain regions implicated in reward and emotion. *Proceedings of the National Academy of Sciences of the United States of America* 98:11818–11823.

Bloom, P. 1996. Intention, history, and artifact concept. *Cognition* 60:1–29.

Bloom, P. 2004. *Descartes' Baby: How Child Development Explains What Makes Us Human*. London: Arrow Books.

Bloom, P. 2007. Religion is natural. *Developmental Science* 10:147–151.

Bloom, P. 2009. Religious belief as an evolutionary accident. In *The Believing Primate: Scientific, Philosophical, and Theological Reflections on the Origin of Religion*, ed. J. Schloss and M. Murray, 118–127. Oxford: Oxford University Press.

Bloom, P., and L. Markson. 1998. Intention and analogy in children's naming of pictorial representations. *Psychological Science* 9:200–204.

Blumstein, D. T. 2007. The evolution of alarm communication in rodents: Structure, function, and the puzzle of apparently altruistic calling in rodents. In *Rodent Societies: An Ecological and Evolutionary Perspective*, ed. J. O. Wolff and P. W. Sherman, 317–327. Chicago: University of Chicago Press.

BonJour, L. 1985. *The Structure of Empirical Knowledge*. Cambridge, MA: Harvard University Press.

Boudry, M., and J. De Smedt. 2011. In mysterious ways: On the modus operandi of supernatural beings. *Religion* 41:449–469.

Bowler, P. J. 2007. *Monkey Trials and Gorilla Sermons: Evolution and Christianity from Darwin to Intelligent Design*. Cambridge, MA: Harvard University Press.

Bowles, S., and H. Gintis. 2011. *A Cooperative Species: Human Reciprocity and Its Evolution*. Princeton: Princeton University Press.

Boyer, P. 1994. *The Naturalness of Religious Ideas: A Cognitive Theory of Religion*. Berkeley: University of California Press.

Boyer, P. 2002. *Religion Explained: The Evolutionary Origins of Religious Thought*. London: Vintage.

Boyer, P., and B. Bergstrom. 2008. Evolutionary perspectives on religion. *Annual Review of Anthropology* 37:111–130.

Boyer, P., and C. Ramble. 2001. Cognitive templates for religious concepts: Cross-cultural evidence for recall of counter-intuitive representations. *Cognitive Science* 25:535–564.

Bräuer, J., J. Kaminski, J. Riedel, J. Call, and M. Tomasello. 2006. Making inferences about the location of hidden food: Social dog, causal ape. *Journal of Comparative Psychology* 120:38–46.

Brewer, W., C. A. Chinn, and A. Samarapungavan. 2000. Explanation in scientists and children. In *Explanation and Cognition*, ed. F. C. Keil and R. A. Wilson, 279–298. Cambridge, MA: MIT Press.

Brown, C. M. 2008. The design argument in classical Hindu thought. *Journal of Hindu Studies* 12:103–151.

Brown, C. M. 2012. *Hindu Perspectives on Evolution: Darwin, Dharma, and Design.* London: Routledge.

Brown, D. E. 1991. *Human Universals.* New York: McGraw-Hill.

Brown, S., X. Gao, L. Tisdelle, S. B. Eickhoff, and M. Liotti. 2011. Naturalizing aesthetics: Brain areas for aesthetic appraisal across sensory modalities. *NeuroImage* 58:250–258.

Buffon, G.-L. L. 1766. *Histoire naturelle générale et particulière: Avec la description du cabinet du roi* (vol. 14). Paris: Imprimerie Royale.

Bullot, N. J., and R. Reber. 2013. The artful mind meets art history: Toward a psycho-historical framework for the science of art appreciation. *Behavioral and Brain Sciences* 36:123–180.

Burge, T. 1993. Content preservation. *Philosophical Review* 102:457–488.

Burke, E. 1757. *A Philosophical Enquiry into the Origin of Our Ideas of the Sublime and Beautiful.* London: R. and J. Dodsley.

Burns, R. M. 1981. *The Great Debate on Miracles: From Joseph Glanvill to David Hume.* Lewisburg, PA: Bucknell University Press.

Butler, J. 1736. *The Analogy of Religion, Natural and Revealed, to the Constitution and Course of Nature.* London: James, John, and Paul Knapton.

Caldwell-Harris, C. L., A. L. Wilson, E. LoTempio, and B. Beit-Hallahmi. 2011. Exploring the atheist personality: Well-being, awe, and magical thinking in atheists, Buddhists, and Christians. *Mental Health, Religion & Culture* 14:659–672.

Call, J., and M. Tomasello. 1999. A non-verbal false belief task: The performance of children and great apes. *Child Development* 70:381–395.

Call, J., and M. Tomasello. 2008. Does the chimpanzee have a theory of mind? 30 years later. *Trends in Cognitive Sciences* 12:187–192.

Callaghan, T., P. Rochat, A. Lillard, M. L. Claux, H. Odden, S. Itakura, S. Tapanya, and S. Singh. 2005. Synchrony in the onset of mental-state reasoning: Evidence from five cultures. *Psychological Science* 16:378–384.

Calvin, J. (1559) 1960. *Institutes of the Christian Religion.* Trans. F. L. Battles. Philadelphia: Westminster Press.

Caramazza, A., and B. Z. Mahon. 2003. The organization of conceptual knowledge: The evidence from category-specific deficits. *Trends in Cognitive Sciences* 7:354–361.

Caramazza, A., and J. R. Shelton. 1998. Domain-specific knowledge systems in the brain: The animate-inanimate distinction. *Journal of Cognitive Neuroscience* 10: 1–34.

Carey, S., and E. S. Spelke. 1996. Science and core knowledge. *Philosophy of Science* 63:515–533.

Carpenter, S. 2012. Government sanctions Harvard psychologist. *Science* 337:1283.

Carroll, N. 1993. On being moved by nature: Between religion and natural history. In *Landscape, Natural Beauty, and the Arts,* ed. S. Kemal and I. Gaskell, 224–266. Cambridge: Cambridge University Press.

Carruthers, P. 2006. *The Architecture of the Mind.* Oxford: Clarendon Press.

Cashdan, E. A. 1980. Egalitarianism among hunters and gatherers. *American Anthropologist* 82:116–120.

Casler, K., and D. Kelemen. 2005. Young children's rapid learning about artifacts. *Developmental Science* 8:472–480.

Casler, K., and D. Kelemen. 2007. Reasoning about artifacts at 24 months: The developing teleo-functional stance. *Cognition* 103:120–130.

Casler, K., and D. Kelemen. 2008. Developmental continuity in teleo-functional explanation: Reasoning about nature among Romanian Romani adults. *Journal of Cognition and Development* 9:340–362.

Cavanagh, P. 2005. The artist as neuroscientist. *Nature* 434:301–307.

Cicero, M. T. (45 BCE) 1967. The nature of the gods. In *Cicero in Twenty-Eight Volumes* (vol. 19), trans. H. Rackham, 2–383. London: William Heineman.

Clark, K. J. 1990. *Return to Reason: A Critique of Enlightenment Evidentialism and a Defense of Reason and Belief in God.* Grand Rapids, MI: Eerdmans.

Clark, K. J., and J. L. Barrett. 2010. Reformed epistemology and the cognitive science of religion. *Faith and Philosophy* 27:174–189.

Clark, K. J., and J. L. Barrett. 2011. Reidian religious epistemology and the cognitive science of religion. *Journal of the American Academy of Religion* 79:639–675.

Clarke, S. 1728. *A Discourse Concerning the Being and Attributes of God, the Obligations of Natural Religion, and the Truth and Certainty of the Christian Revelation*, 7th ed. London: James and John Knapton.

Clottes, J., M. Menu, and P. Walter. 1990. La préparation des peintures magdaléni-ennes des cavernes ariégeoises. *Bulletin de la Société Préhistorique française* 87: 170–192.

Collins, F. 2007. *The Language of God: A Scientist Presents Evidence for Belief*. London: Simon & Schuster.

Comte, A. 1841. *Cours de philosophie positive: La partie historique de la philosophie sociale en tout ce qui concerne l'état théologique et l'état métaphysique*, vol. 5. Paris: Bachelier.

Conard, N. J. 2003. Palaeolithic ivory sculptures from southwestern Germany and the origins of figurative art. *Nature* 426:830–832.

Conard, N. J., M. Malina, and S. C. Münzel. 2009. New flutes document the earliest musical tradition in southwestern Germany. *Nature* 460:737–740.

Conee, E., and E. Feldman. 1998. The generality problem for reliabilism. *Philosophical Studies* 89:1–29.

Conway Morris, S. 2003. *Life's Solution: Inevitable Humans in a Lonely Universe*. Cambridge: Cambridge University Press.

Copan, P. 2008. The moral argument. In *Philosophy of Religion: Classic and Contemporary Issues*, ed. P. Copan and C. Meister, 127–141. Malden, MA: Blackwell.

Copp, D. 2008. Darwinian skepticism about moral realism. *Philosophical Issues* 18:186–206.

Corbin, A. 1988. *Le territoire du vide: L'occident et le désir du rivage (1750–1840)*. Paris: Aubier.

Cordry, B. S. 2011. A more dangerous enemy? Philo's "confession" and Hume's soft atheism. *International Journal for Philosophy of Religion* 70:61–83.

Corriveau, K. H., E. E. Chen, and P. L. Harris. In press. Judgments about fact and fiction by children from religious and non-religious backgrounds. *Cognitive Science*.

Cosmides, L., and J. Tooby. 1994a. Beyond intuition and instinct blindness: Toward an evolutionarily rigorous cognitive science. *Cognition* 50:41–77.

Cosmides, L., and J. Tooby. 1994b. Origins of domain specificity: The evolution of functional organization. In *Mapping the Mind: Domain Specificity in Cognition and Culture*, ed. L. Hirschfeld and S. A. Gelman, 85–116. Cambridge: Cambridge University Press.

Cottingham, J. 2012. Human nature and the transcendent. *Royal Institute of Philosophy* 70 (Supplement):233–254.

Craffert, P. F. 2009. Jesus' resurrection in a social-scientific perspective: Is there anything new to be said? *Journal for the Study of the Historical Jesus* 7:126–151.

Craig, W. L. 2001. Prof. Grünbaum on the "normalcy of nothingness" in the Leibnizian and kalam cosmological arguments. *British Journal for the Philosophy of Science* 52:371–386.

Craig, W. L. 2003. The cosmological argument. In *The Rationality of Theism*, ed. P. Copan and P. K. Moser, 112–131. London: Routledge.

Craig, W. L., and J. P. Moreland, eds. 2009. *Blackwell Companion to Natural Theology*. Malden, MA: Wiley Blackwell.

Crossley, J. G. 2006. *Why Christianity Happened: A Sociohistorical Account of Christian Origins (26–50 CE)*. Louisville, KY: Westminster John Knox Press.

Cummins, R. 2002. Neo-teleology. In *Functions: New Essays in the Philosophy of Psychology and Biology*, ed. A. Ariew, R. Cummins, and M. Perlman, 157–172. New York: Oxford University Press.

Dalla Bella, S., J. Giguère, and I. Peretz. 2007. Singing proficiency in the general population. *Journal of the Acoustical Society of America* 121:1182–1189.

Darwin, C. 1859. *On the Origin of Species by Means of Natural Selection or the Preservation of Favoured Races in the Struggle for Life*. London: John Murray.

Darwin, C. 1871a. *The Descent of Man, and Selection in Relation to Sex*, vol. 1. London: John Murray.

Darwin, C. 1871b. *The Descent of Man, and Selection in Relation to Sex*, vol. 2. London: John Murray.

Dasti, M. R. 2011. Indian rational theology: Proof, justification, and epistemic liberality in Nyāya's argument for God. *Asian Philosophy: An International Journal of the Philosophical Traditions of the East* 21:1–21.

Davidson, D. 1963. Actions, reasons, and causes. *Journal of Philosophy* 60:685–700.

Davies, S. 2012. *The Artful Species: Aesthetics, Art, and Evolution*. Oxford: Oxford University Press.

Dawid, R., S. Hartmann, and J. Sprenger. In press. The *no alternatives* argument. *British Journal for the Philosophy of Science*.

Dawkins, R. 1986. *The Blind Watchmaker*. London: Penguin.

Dawkins, R. 2006. *The God Delusion*. Boston: Houghton Mifflin.

Dean, L. G., R. L. Kendal, S. J. Schapiro, B. Thierry, and K. N. Laland. 2012. Identification of the social and cognitive processes underlying human cumulative culture. *Science* 335:1114–1118.

De Cruz, H. In press. Cognitive science of religion and the study of theological concepts. *Topoi.*

De Cruz, H. 2014. The enduring appeal of natural theological arguments. *Philosophy Compass* 9:145–153.

De Cruz, H., M. Boudry, J. De Smedt, and S. Blancke. 2011. Evolutionary approaches to epistemic justification. *Dialectica* 65:517–535.

De Cruz, H., and J. De Smedt. 2007. The role of intuitive ontologies in scientific understanding—The case of human evolution. *Biology and Philosophy* 22:351–368.

De Cruz, H., and J. De Smedt. 2010a. Paley's iPod: The cognitive basis of the design argument within natural theology. *Zygon: Journal of Religion and Science* 45:665–684.

De Cruz, H., and J. De Smedt. 2010b. Science as structured imagination. *Journal of Creative Behavior* 44:29–44.

De Cruz, H., and J. De Smedt. 2012. Evolved cognitive biases and the epistemic status of scientific beliefs. *Philosophical Studies* 157:411–429.

De Cruz, H., and J. De Smedt. 2013. Reformed and evolutionary epistemology and the noetic effects of sin. *International Journal for Philosophy of Religion* 74:49–66.

de Gardelle, V., and C. Summerfield. 2011. Robust averaging during perceptual judgment. *Proceedings of the National Academy of Sciences of the United States of America* 108:13341–13346.

Dehaene, S., and L. Cohen. 2007. Cultural recycling of cortical maps. *Neuron* 56:384–398.

Dembski, W. A. 1998. *The Design Inference: Eliminating Chance through Small Probabilities.* Cambridge: Cambridge University Press.

Dembski, W. A. 1999. *Intelligent Design: The Bridge between Science and Theology.* Downers Grove, IL: InterVarsity Press.

Dennett, D. C. 1987. *The Intentional Stance.* Cambridge, MA: MIT Press.

Dennett, D. C. 2006. *Breaking the Spell: Religion as a Natural Phenomenon.* Oxford: Allen Lane.

Descartes, R. (1619) 1985. Rules for the direction of the mind. In *The Philosophical Writings of Descartes*, vol. 1, trans. J. Cottingham, R. Stoothoff, and D. Murdoch, 9–78. Cambridge: Cambridge University Press.

De Smedt, J., and H. De Cruz. 2010. Toward an integrative approach of cognitive neuroscientific and evolutionary psychological studies of art. *Evolutionary Psychology* 8:695–719.

De Smedt, J., and H. De Cruz. 2011. The cognitive appeal of the cosmological argument. *Method and Theory in the Study of Religion* 23:103–122.

De Smedt, J., and H. De Cruz. 2012. Human artistic behaviour: Adaptation, byproduct, or cultural group selection? In *Philosophy of Behavioral Biology*, ed. K. Plaisance and T. Reydon, 167–187. Dordrecht: Springer.

De Smedt, J., and H. De Cruz. 2013a. The artistic design stance and the interpretation of Paleolithic art. *Behavioral and Brain Sciences* 36:139–140.

De Smedt, J., and H. De Cruz. 2013b. Delighting in natural beauty: Joint attention and the phenomenology of nature aesthetics. *European Journal for Philosophy of Religion* 5:167–186.

De Smedt, J., and H. De Cruz. 2014. The *imago Dei* as a work in progress: A perspective from paleoanthropology. *Zygon: Journal of Religion and Science* 49:135–156.

De Smedt, J., H. De Cruz, and J. Braeckman. 2009. Why the human brain is not an enlarged chimpanzee brain. In *Human Characteristics: Evolutionary Perspectives on Human Mind and Kind*, ed. H. Høgh-Olesen, J. Tønnesvang, and P. Bertelsen, 168–181. Newcastle upon Tyne: Cambridge Scholars.

de Waal, F. 2009. *The Age of Empathy: Nature's Lessons for a Kinder Society*. New York: Random House.

Dewey, J. 1898. Evolution and ethics. *Monist* 8:321–341.

Diodorus Siculus. (1st c. BCE) 1939. *Bibliotheca Historica* (vol. 3). Trans. C. H. Oldfather. Cambridge, MA: Harvard University Press.

Dissanayake, E. 2000. *Art and Intimacy: How the Arts Began*. Washington, D.C.: University of Washington Press.

Dobzhansky, T. 1973. Nothing in biology makes sense except in the light of evolution. *American Biology Teacher* 35:125–129.

Douven, I. 2002. Testing inference to the best explanation. *Synthese* 130:355–377.

Dubreuil, B. 2010. Paleolithic public goods games: Why human culture and cooperation did not evolve in one step. *Biology and Philosophy* 25:53–73.

Durkheim, E. 1915. *The Elementary Forms of the Religious Life: A Study in Religious Sociology*. Trans. J. W. Swain. London: Allen & Unwin.

Edwards, L. C. 2012. Re-envisaging Ruskin's types: Beautiful order as divine revelation. *Irish Theological Quarterly* 77:165–181.

El-Bizri, N. 2008. God: Essence and attributes. In *The Cambridge Companion to Classical Islamic Theology*, ed. T. Winter, 121–140. Cambridge: Cambridge University Press.

Eshleman, A. S. 2005. Can an atheist believe in God? *Religious Studies* 41:183–199.

Evans, C. S. 2010. *Natural Signs and Knowledge of God: A New Look at Theistic Arguments*. Oxford: Oxford University Press.

Evans, E. M. 2001. Cognitive and contextual factors in the emergence of diverse belief systems: Creation versus evolution. *Cognitive Psychology* 42:217–266.

Evans, J. S. B. T. 2008. Dual-processing accounts of reasoning, judgment and social cognition. *Annual Review of Psychology* 59:255–278.

Fales, E. 1996. Plantinga's case against naturalistic epistemology. *Philosophy of Science* 63:432–451.

Falk, J. H., and J. D. Balling. 2010. Evolutionary influence on human landscape preference. *Environment and Behavior* 42:479–493.

Farroni, T., G. Csibra, F. Simion, and M. H. Johnson. 2002. Eye contact detection in humans from birth. *Proceedings of the National Academy of Sciences of the United States of America* 99:9602–9605.

Fastl, H., and E. Zwicker. 2007. *Psychoacoustics: Facts and Models*. Berlin: Springer.

Faust, J. 2008. Can religious arguments persuade? *International Journal for Philosophy of Religion* 63:71–86.

Fichman, M. 2001. Science in theistic contexts: A case study of Alfred Russel Wallace on human evolution. *Osiris* 16:227–250.

Flew, A. 1985. Introduction to *An Enquiry Concerning Human Understanding*, by David Hume, vii–xx. Peru, IL: Open Court.

Fondevila, S., and M. Martín-Loeches. 2013. Cognitive mechanisms for the evolution of religious thought. *Annals of the New York Academy of Sciences* 1299:84–90.

Fondevila, S., M. Martín-Loeches, L. Jiménez-Ortega, P. Casado, A. Sel, A. Fernández-Hernández, and W. Sommer. 2012. The sacred and the absurd—an electrophysiological study of counterintuitive ideas (at sentence level). *Social Neuroscience* 7:445–457.

Forrest, B., and P. R. Gross. 2004. *Creationism's Trojan Horse: The Wedge of Intelligent Design*. Oxford: Oxford University Press.

Frank, P. 2004. On the assumption of design. *Theology and Science* 2:109–130.

Freud, S. 1927. *Die Zukunft einer Illusion*. Leipzig: Internationaler Psychoanalytischer Verlag.

Fricker, E. 1994. Against gullibility. In *Similarity and Analogical Reasoning*, ed. B. K. Matilal and A. Chakrabarti, 125–161. Boston: Kluwer.

Fricker, E. 2006. Second-hand knowledge. *Philosophy and Phenomenological Research* 73:592–618.

Frith, C. D., and U. Frith. 1999. Interacting minds—A biological basis. *Science* 286:1692–1695.

Frumkin, H. 2001. Beyond toxicity: Human health and the natural environment. *American Journal of Preventive Medicine* 20:234–240.

Garwood, C. 2008. *Flat Earth: The History of an Infamous Idea.* London: Pan Macmillan.

Gelman, S. A., and P. Bloom. 2000. Young children are sensitive to how an object was created when deciding what to name it. *Cognition* 76:91–103.

Gelman, S. A., and K. S. Ebeling. 1998. Shape and representational status in children's early naming. *Cognition* 66:B35–B47.

Gelman, S. A., and G. Gottfried. 1996. Children's causal explanations of animate and inanimate motion. *Child Development* 67:1970–1987.

Gelman, S. A., and H. M. Wellman. 1991. Insides and essences: Early understandings of the non-obvious. *Cognition* 38:213–244.

Gentner, D., S. Brem, R. W. Ferguson, A. B. Markman, B. B. Levidow, P. Wolff, and K. D. Forbus. 1997. Analogical reasoning and conceptual change: A case study of Johannes Kepler. *Journal of the Learning Sciences* 6:3–40.

Gergely, G., Z. Nádasdy, G. Csibra, and S. Bíró. 1995. Taking the intentional stance at 12 months of age. *Cognition* 56:165–193.

German, T. P., and H. C. Barrett. 2005. Functional fixedness in a technologically sparse culture. *Psychological Science* 16:1–5.

German, T. P., and S. C. Johnson. 2002. Function and the origins of the design stance. *Journal of Cognition and Development* 3:279–300.

Gervais, W. M., and J. Henrich. 2010. The Zeus problem: Why representational content biases cannot explain faith in gods. *Journal of Cognition and Culture* 10:383–389.

Gervais, W. M., A. K. Willard, A. Norenzayan, and J. Henrich. 2011. The cultural transmission of faith: Why innate intuitions are necessary, but insufficient, to explain religious belief. *Religion* 41:389–410.

Gill, K. Z., and D. Purves. 2009. A biological rationale for musical scales. *PLoS ONE* 4:e8144.

Gintis, H., E. A. Smith, and S. Bowles. 2001. Costly signaling and cooperation. *Journal of Theoretical Biology* 213:103–119.

Gleeson, A. 2010. The power of God. *Sophia* 49:603–616.

Gliboff, S. 2000. Paley's design argument as an inference to the best explanation, or, Dawkins' dilemma. *Studies in History and Philosophy of Science C* 31:579–597.

Godfrey-Smith, P. 1991. Signal, decision, action. *Journal of Philosophy* 88:709–722.

Goldman, A. I. 1994. Naturalistic epistemology and reliabilism. *Midwest Studies in Philosophy* 19:301–320.

Goldman, A. I. 1999. A priori warrant and naturalistic epistemology. In *Philosophical Perspectives 13: Epistemology*, ed. J. Tomberlin, 1–28. Oxford: Blackwell.

Goldman, A. I. 2001. Experts: Which ones should you trust? *Philosophy and Phenomenological Research* 63:85–110.

Goodnick, L. 2012. Cleanthes's propensity and intelligent design. *Modern Schoolman* 88:299–316.

Goodwin, G. P., and J. M. Darley. 2008. The psychology of meta-ethics: Exploring objectivism. *Cognition* 106:1339–1366.

Gopnik, A., and A. Meltzoff. 1997. *Words, Thoughts, and Theories*. Cambridge, MA: MIT Press.

Gray, A. 1888. *Darwiniana: Essays and Reviews Pertaining to Darwinism*. New York: Appleton.

Greenblatt, S. H. 1995. Phrenology in the science and culture of the 19th century. *Neurosurgery* 37:790–805.

Griffiths, T. L., and J. B. Tenenbaum. 2007. From mere coincidences to meaningful discoveries. *Cognition* 103:180–226.

Grünbaum, A. 2000. A new critique of theological interpretations of physical cosmology. *British Journal for the Philosophy of Science* 51:1–43.

Guthrie, S. E. 1993. *Faces in the Clouds: A New Theory of Religion*. New York: Oxford University Press.

Habermas, G. R. 2005. Resurrection research from 1975 to the present: What are critical scholars saying? *Journal for the Study of the Historical Jesus* 3:135–153.

Haidt, J. 2001. The emotional dog and its rational tail: A social intuitionist approach to moral judgment. *Psychological Review* 108:814–834.

Hájek, A. 2008. Are miracles chimerical? In *Oxford Studies in Philosophy of Religion*, vol. 1, ed. J. Kvanvig, 82–104. Oxford: Oxford University Press.

Hall, J. W. 2009. Chance for a purpose. *Perspectives on Science and Christian Faith* 61:3–11.

Hamilton, W. D. 1963. The evolution of altruistic behavior. *American Naturalist* 97:354–356.

Hamlin, J. K., K. Wynn, and P. Bloom. 2007. Social evaluation by preverbal infants. *Nature* 450:557–559.

Hamlin, J. K., K. Wynn, P. Bloom, and N. Mahajan. 2011. How infants and toddlers react to antisocial others. *Proceedings of the National Academy of Sciences of the United States of America* 108:19931–19936.

Hardy, L. 2012. Hume's defense of true religion. In *The Persistence of the Sacred in Modern Thought*, ed. C. L. Firestone and N. A. Jacobs, 251–272. Notre Dame, IN: University of Notre Dame Press.

Hare, B., J. Call, and M. Tomasello. 2001. Do chimpanzees know what conspecifics know? *Animal Behaviour* 61:139–151.

Hare, J. 2004. Is there an evolutionary foundation for human morality? In *Evolution and Ethics: Human Morality in Biological and Religious Perspective*, ed. P. Clayton and J. Schloss, 187–203. Grand Rapids, MI: Wm. B. Eerdmans.

Harris, P. L., and K. H. Corriveau. 2014. Learning from testimony about religion and science. In *Trust and Skepticism. Children's Selective Learning from Testimony*, ed. E. J. Robinson and S. Einav, 28–41. Hove: Psychology Press.

Harris, P. L., and M. Giménez. 2005. Children's acceptance of conflicting testimony: The case of death. *Journal of Cognition and Culture* 5:143–164.

Harris, P. L., E. S. Pasquini, S. Duke, J. J. Asscher, and F. Pons. 2006. Germs and angels: The role of testimony in young children's ontology. *Developmental Science* 9:76–96.

Harrison, P. 1995. Newtonian science, miracles, and the laws of nature. *Journal of the History of Ideas* 56:531–553.

Hasker, W. 2007. D.Z. Phillips' problems with evil and with God. *International Journal for Philosophy of Religion* 61:151–160.

Hassin, R. R., J. A. Bargh, and J. S. Uleman. 2002. Spontaneous causal inferences. *Journal of Experimental Social Psychology* 38:515–522.

Haught, J. F. 2000. *God after Darwin: A Theology of Evolution*. Boulder, CO: Westview Press.

Henrich, J. 2004. Cultural group selection, coevolutionary processes and large-scale cooperation. *Journal of Economic Behavior & Organization* 53:3–35.

Henrich, J., S. J. Heine, and A. Norenzayan. 2010. The weirdest people in the world? *Behavioral and Brain Sciences* 3:61–83.

Henrich, J., R. McElreath, A. Barr, J. Ensminger, C. Barrett, A. Bolyanatz, J. C. Cardenas, et al. 2006. Costly punishment across human societies. *Science* 312: 1767–1770.

Heschel, A. J. [1955] 2009. *God in Search of Man: A Philosophy of Judaism*. London: Souvenir.

Heschel, A. J. 1965. *Who Is Man?* Stanford, CA: Stanford University Press.

Himma, K. E. 2005. The application-conditions for design inferences: Why the design arguments need the help of other arguments for God's existence. *International Journal for Philosophy of Religion* 57:1–33.

Hodge, K. M. 2008. Descartes' mistake: How afterlife beliefs challenge the assumption that humans are intuitive Cartesian substance dualists. *Journal of Cognition and Culture* 8:387–415.

Holland, R. F. 1965. The miraculous. *American Philosophical Quarterly* 2:43–51.

Horner, V., and A. Whiten. 2005. Causal knowledge and imitation/emulation switching in chimpanzees (*Pan troglodytes*) and children (*Homo sapiens*). *Animal Cognition* 8:164–181.

Horst, S. 2013. Notions of intuition in the cognitive science of religion. *Monist* 96:377–398.

Hoyle on evolution. 1981. *Nature* 294:105.

Hume, D. 1748. *Philosophical Essays Concerning Human Understanding*. London: A. Millar.

Hume, D. 1757. The natural history of religion. In *Four Dissertations*, 1–117. London: A. Millar.

Hume, D. 1779. *Dialogues Concerning Natural Religion*, 2nd ed. London: Hafner.

Hurley, S. 2003. Animal action in the space of reasons. *Mind and Language* 18:231–257.

Huron, D. 2006. *Sweet Anticipation: Music and the Psychology of Expectation*. Cambridge, MA: MIT Press.

Itakura, S. 2004. Gaze-following and joint visual attention in nonhuman animals. *Japanese Psychological Research* 46:216–226.

Ivey, P. K. 2000. Cooperative reproduction in Ituri forest hunter-gatherers: Who cares for Efe infants? *Current Anthropology* 41:856–866.

Jackson, P. L., A. N. Meltzoff, and J. Decety. 2005. How do we perceive the pain of others? A window into the neural processes involved in empathy. *NeuroImage* 24:771–779.

James, S. M. 2011. *An Introduction to Evolutionary Ethics*. Chichester: Wiley-Blackwell.

James, W. 1902. *The Varieties of Religious Experience: A Study in Human Nature*. New York: Longmans, Green.

Jaswal, V. K. 2006. Preschoolers favor the creator's label when reasoning about an artifact's function. *Cognition* 99:B83–B92.

Jim, C. Y., and Y. Chen. 2009. Value of scenic views: Hedonic assessment of private housing in Hong Kong. *Landscape and Urban Planning* 91:226–234.

Johnson, J. L., and J. Potter. 2005. The argument from language and the existence of God. *Journal of Religion* 85:83–93.

Johnson, P. E. 2000. *The Wedge of Truth: Splitting the Foundations of Naturalism*. Downers Grove, IL: InterVarsity Press.

Joyce, R. 2006. *The Evolution of Morality*. Cambridge, MA: MIT Press.

Kahane, G. 2011. Evolutionary debunking arguments. *Noûs* 45:103–125.

Kail, P. J. E. 2007. Understanding Hume's natural history of religion. *Philosophical Quarterly* 57:190–211.

Kaminski, J., J. Call, and M. Tomasello. 2008. Chimpanzees know what others know, but not what they believe. *Cognition* 109:224–234.

Kant, I. (1781) 2005. *Critique of Pure Reason*. Ed. P. Guyer and A. W. Wood. Cambridge: Cambridge University Press.

Kant, I. (1790) 1987. *Critique of the power of judgment*. Ed. P. Guyer, trans. P. Guyer and E. Matthews. Cambridge: Cambridge University Press.

Kapogiannis, D., A. K. Barbey, M. Su, G. Zamboni, F. Krueger, and J. Grafman. 2009. Cognitive and neural foundations of religious belief. *Proceedings of the National Academy of Sciences of the United States of America* 106:4876–4881.

Keener, C. S. 2011a. Assumptions in historical-Jesus research: Using ancient biographies and disciples' traditioning as a control. *Journal for the Study of the Historical Jesus* 9:26–58.

Keener, C. S. 2011b. *Miracles: The Credibility of the New Testament Accounts*. Grand Rapids, MI: Baker Academic.

Keil, F. C. 1989. *Concepts, Kinds, and Cognitive Development*. Cambridge, MA: MIT Press.

Keil, F. C. 2003. Folkscience: Coarse interpretations of a complex reality. *Trends in Cognitive Sciences* 7:368–373.

Kelemen, D. 2003. British and American children's preferences for teleo-functional explanations of the natural world. *Cognition* 88:201–221.

Kelemen, D. 2004. Are children "intuitive theists"? Reasoning about purpose and design in nature. *Psychological Science* 15:295–301.

Kelemen, D., and S. A. Carey. 2007. The essence of artifacts: Developing the design stance. In *Creations of the Mind: Theories of Artifacts and Their Representation*, ed. E. Margolis and S. Laurence, 212–230. Oxford: Oxford University Press.

Kelemen, D., and C. DiYanni. 2005. Intuitions about origins: Purpose and intelligent design in children's reasoning about nature. *Journal of Cognition and Development* 6:3–31.

Kelemen, D., and E. Rosset. 2009. The human function compunction: Teleological explanation in adults. *Cognition* 111:138–143.

Kelemen, D., J. Rottman, and R. Seston. 2013. Professional physical scientists display tenacious teleological tendencies: Purpose-based reasoning as a cognitive default. *Journal of Experimental Psychology: General* 142:1074–1083.

Kelemen, D., D. Widdowson, T. Posner, A. Brown, and K. Casler. 2003. Teleo-functional constraints on preschool children's reasoning about living things. *Developmental Science* 6:329–345.

Kelly, T. 2008. Disagreement, dogmatism, and belief polarization. *Journal of Philosophy* 105:611–633.

Kelly, T. 2011. *Consensus gentium*: Reflections on the "common consent" argument for the existence of God. In *Evidence and Religious Belief*, ed. K. J. Clark and R. J. VanArragon, 135–156. Oxford: Oxford University Press.

Keltner, D., and J. Haidt. 2003. Approaching awe, a moral, spiritual, and aesthetic emotion. *Cognition and Emotion* 17:297–314.

Kennedy, G. E. 2005. From the ape's dilemma to the weanling's dilemma: Early weaning and its evolutionary context. *Journal of Human Evolution* 48:123–145.

Keysar, B., D. J. Barr, J. A. Balin, and J. S. Brauner. 2000. Taking perspective in conversation: The role of mutual knowledge in comprehension. *Psychological Science* 11:32–38.

Knight, N., P. Sousa, J. L. Barrett, and S. Atran. 2004. Children's attributions of beliefs to humans and God: Cross-cultural evidence. *Cognitive Science* 28:117–126.

Koenig, M. A., and P. L. Harris. 2007. The basis of epistemic trust: Reliable testimony or reliable sources? *Episteme* 4:264–283.

Kohák, E. 1984. *The Embers and the Stars: A Philosophical Inquiry into the Moral Sense of Nature*. Chicago: University of Chicago Press.

Koons, R. C. 1997. A new look at the cosmological argument. *American Philosophical Quarterly* 34:193–211.

Koppers, W. 1924. *Unter Feuerland-Indianern: Eine Forschungsreise zu den südlichsten Bewohnern der Erde*. Stuttgart: Strecker und Schröder.

Kornblith, H. 2010. What reflective endorsement cannot do. *Philosophy and Phenomenological Research* 80:1–19.

Koslowski, B., J. Marasia, M. Chelenza, and R. Dublin. 2008. Information becomes evidence when an explanation can incorporate it into a causal framework. *Cognitive Development* 23:472–487.

Kovács, Á. M., E. Téglás, and A. D. Endress. 2010. The social sense: Susceptibility to others' beliefs in human infants and adults. *Science* 330:1830–1834.

Kozhevnikov, M., and M. Hegarty. 2001. Impetus beliefs as default heuristics: Dissociation between explicit and implicit knowledge about motion. *Psychonomic Bulletin & Review* 8:439–453.

Krams, I., T. Krama, K. Igaune, and R. Mänd. 2008. Experimental evidence of reciprocal altruism in the pied flycatcher. *Behavioral Ecology and Sociobiology* 62:599–605.

Kuhlmeier, V., P. Bloom, and K. Wynn. 2004. Do 5-month-old infants see humans as material objects? *Cognition* 94:95–103.

Kuzmin, Y. V., G. S. Burr, A. J. T. Jull, and L. D. Sulerzhitsky. 2004. AMS ^{14}C age of the Upper Palaeolithic skeletons from Sungir site, Central Russian Plain. *Nuclear Instruments and Methods in Physics Research: Section B, Beam Interactions with Materials and Atoms* 223:731–734.

Lahti, D. 2003. Parting with illusions in evolutionary ethics. *Biology and Philosophy* 18:639–651.

Lamarck, J.-B. 1809. *Philosophie zoologique, ou exposition des considérations relatives à l'histoire naturelle des animaux*. Paris: Duminil-Lesueur.

Lane, J. D., H. M. Wellman, and E. M. Evans. 2010. Children's understanding of ordinary and extraordinary minds. *Child Development* 81:1475–1489.

Larmer, R. 2002. Is there anything wrong with "God of the gaps" reasoning? *International Journal for Philosophy of Religion* 52:129–142.

Larson, J. L. 1979. Vital forces: Regulative principles or constitutive agents? A strategy in German physiology, 1786–1802. *Isis* 70:235–249.

Law, S. 2011. Evidence, miracles, and the existence of Jesus. *Faith and Philosophy* 28:129–151.

Legare, C. H., E. M. Evans, K. S. Rosengren, and P. L. Harris. 2012. The coexistence of natural and supernatural explanations across cultures and development. *Child Development* 83:779–793.

Lehrer, K. 2006. Knowing content in the visual arts. In *Knowing Art: Essays in Aesthetics and Epistemology*, ed. M. Kieran and D. McIver Lopes, 1–26. Dordrecht: Springer.

Leibniz, G. (1714) 1898. The monadology. In *The Monadology and Other Philosophical Writings*, trans. R. Latta, 215–271. London: Oxford University Press.

Le Poidevin, R. 1996. *Arguing for Atheism: An Introduction to the Philosophy of Religion.* London: Routledge.

Lewis, C. S. (1949) 2001. The weight of glory. In *The Weight of Glory and Other Addresses*, 25–46. New York: HarperCollins.

Lewis, C. S. (1952) 2002. *Mere Christianity.* London: Harper Collins.

Lindsay, D. S. 2008. Source monitoring. In *Learning and Memory: A Comprehensive Reference*, vol. 2, ed. H. L. Roediger, 325–348. Oxford: Elsevier.

Littlejohn, C. 2012. *Justification and the Truth-Connection.* Cambridge: Cambridge University Press.

Locke, J. 1690. *An Essay Concerning Human Understanding.* London: Thomas Basset.

Loftus, E. 2003. Make-believe memories. *American Psychologist* 58:867–873.

Lombrozo, T., D. Kelemen, and D. Zaitchik. 2007. Inferring design: Evidence of a preference for teleological explanations in patients with Alzheimer's disease. *Psychological Science* 18:999–1006.

Lucretius. (ca. 50 BCE) 2007. *The Nature of Things (De rerum natura).* Trans. A. E. Stallings. London: Penguin.

Luhrmann, T. M. 2012. *When God Talks Back: Understanding the American Evangelical Relationship with God.* New York: Vintage.

Lutz, D. J., and F. C. Keil. 2002. Early understanding of the division of cognitive labor. *Child Development* 73:1073–1084.

Mackie, J. 1977. *Ethics: Inventing Right and Wrong.* London: Penguin.

Mackie, J. 1982. *The Miracle of Theism: Arguments For and Against the Existence of God.* Oxford: Clarendon Press.

Maimonides, M. (12th c.) 1910. *The Guide for the Perplexed*, 2nd ed. Trans. M. Friedländer. London: George Routledge and Sons.

Malinowski, B. (1925) 1992. *Magic, Science, and Religion and Other Essays.* Prospect Heights, IL: Waveland Press.

Manson, N. A., ed. 2003. *God and Design: The Teleological Argument and Modern Science.* London: Routledge.

Martin, A., and L. R. Santos. 2014. The origins of belief representation: Monkeys fail to automatically represent others' beliefs. *Cognition* 130:300–308.

Martin, A., and J. Weisberg. 2003. Neural foundations for understanding social and mechanical concepts. *Cognitive Neuropsychology* 20:575–587.

McCauley, R. N. 2000. The naturalness of religion and the unnaturalness of science. In *Explanation and Cognition*, ed. F. C. Keil and R. A. Wilson, 61–85. Cambridge, MA: MIT Press.

McCauley, R. N. 2011. *Why Religion Is Natural and Science Is Not.* Oxford: Oxford University Press.

McCloskey, M., A. Caramazza, and B. Green. 1980. Curvilinear motion in the absence of external forces: Naive beliefs about the motion of objects. *Science* 210:1139–1141.

McCloskey, M., A. Washburn, and L. Felch. 1983. Intuitive physics: The straight-down belief and its origin. *Journal of Experimental Psychology: Learning, Memory, and Cognition* 9:636–649.

McDowell, J. 1996. *Mind and World.* Cambridge, MA: Harvard University Press.

McGrath, A. E. 2011. *Darwinism and the Divine: Evolutionary Thought and Natural Theology.* Malden, MA: Wiley-Blackwell.

McGrew, T., and L. McGrew. 2009. The argument from miracles: A cumulative case for the resurrection of Jesus of Nazareth. In *The Blackwell Companion to Natural Theology*, ed. W. L. Craig and J. P. Moreland, 593–662. Chichester: Wiley-Blackwell.

McKay, R. T., and D. C. Dennett. 2009. The evolution of misbelief. *Behavioral and Brain Sciences* 32:493–510.

Medin, D. L., and A. Ortony. 1989. Psychological essentialism. In *Similarity and Analogical Reasoning*, ed. S. Vosniadou and A. Ortony, 179–195. Cambridge: Cambridge University Press.

Menninghaus, W. 2009. Biology à la mode: Charles Darwin's aesthetics of "ornament." *History and Philosophy of the Life Sciences* 31:263–278.

Mercier, H. 2010. The social origins of folk epistemology. *Review of Philosophy and Psychology* 1:499–514.

Metz, K. E. 1998. Emergent understanding and attribution of randomness: Comparative analysis of the reasoning of primary grade children and undergraduates. *Cognition and Instruction* 16:285–365.

Mill, J. S. 1889. *A System of Logic, Ratiocinative and Inductive, Being a Connected View of the Principles of Evidence, and the Methods of Scientific Investigation.* London: Longmans, Green.

Miller, G. 2000. *The Mating Mind: How Sexual Choice Shaped the Evolution of Human Nature*. London: William Heineman.

Miller, J. D., E. C. Scott, and S. Okamoto. 2006. Public acceptance of evolution. *Science* 313:765–766.

Miller, K. R. (1999) 2007. *Finding Darwin's God: A Scientist's Search for Common Ground between God and Evolution*. New York: Harper.

Moll, J., R. De Oliveira-Souza, and R. Zahn. 2008. The neural basis of moral cognition. *Annals of the New York Academy of Sciences* 1124:161–180.

Mongrain, K. 2011. The eyes of reason: Intelligent design apologetics as the new *preambula fidei? Heythrop Journal* 52:191–210.

Montgomery, D. E., and M. Lightner. 2004. Children's developing understanding of differences between their own intentional action and passive movement. *British Journal of Developmental Psychology* 22:417–438.

Moore, E. 1981. A prison environment's effect on health care service demands. *Journal of Environmental Systems* 11:17–34.

Mountford, B. 2011. *Christian Atheist: Belonging without Believing*. Alresford: O-Books.

Murray, M. 2008. Four arguments that the cognitive psychology of religion undermines the justification of religious belief. In *The Evolution of Religion: Studies, Theories, and Critiques*, ed. J. Bulbulia, R. Sosis, E. Harris, R. Genet, C. Genet, and K. Wyman, 393–398. Santa Margarita, CA: Collins Foundation Press.

Nagel, J. 2012. Intuitions and experiments: A defense of the case method in epistemology. *Philosophy and Phenomenological Research* 85:495–527.

Neill, A., and A. Ridley. 2010. Religious music for godless ears. *Mind* 119:999–1023.

New, J., L. Cosmides, and J. Tooby. 2007. Category-specific attention for animals reflects ancestral priorities, not expertise. *Proceedings of the National Academy of Sciences of the United States of America* 104:16598–16603.

Newberg, A., A. Alavi, M. Baime, M. Pourdehnad, J. Santanna, and E. d'Aquili. 2001. The measurement of regional cerebral blood flow during the complex cognitive task of meditation: A preliminary SPECT study. *Psychiatry Research: Neuroimaging Section* 106:113–122.

Newman, G., F. Keil, V. Kuhlmeier, and K. Wynn. 2010. Early understandings of the link between agents and order. *Proceedings of the National Academy of Sciences of the United States of America* 107:17140–17145.

Newman, J. H. (1870) 1973. Newman to William Robert Brownlow, April 13, 1870. In *The Letters and Diaries of John Henry Newman*, vol. 25, ed. C. S. Dessain and T. Gornall, 97. Oxford: Clarendon Press.

Newton, A. M., and J. G. de Villiers. 2007. Thinking while talking: Adults fail non-verbal false-belief reasoning. *Psychological Science* 18:574–579.

Nichols, S. 2004. *Sentimental Rules: On the Natural Foundations of Moral Judgment.* New York: Oxford University Press.

Nichols, S., and T. Folds-Bennett. 2003. Are children moral objectivists? Children's judgments about moral and response-dependent properties. *Cognition* 90:B23–B32.

Nichols, S., S. Stich, and J. M. Weinberg. 2003. Meta-skepticism: Meditations in ethno-epistemology. In *The Skeptics: Contemporary Essays*, ed. S. Luper, 227–247. Aldershot: Ashgate.

Nietzsche, F. (1889) 2005. Twilight of the idols, or How to philosophize with a hammer. In *The Anti-Christ, Ecce homo, Twilight of the Idols, and Other Writings*, ed. A. Ridley and J. Norman, trans. J. Norman, 153–229. Cambridge: Cambridge University Press.

Nieuwland, M. S., and J. J. A. Van Berkum. 2006. When peanuts fall in love: N400 evidence for the power of discourse. *Journal of Cognitive Neuroscience* 18: 1098–1111.

Norenzayan, A., S. Atran, J. Faulkner, and M. Schaller. 2006. Memory and mystery: The cultural selection of minimally counterintuitive narratives. *Cognitive Science* 30:531–553.

Norton, J. D. 2007. Causation as folk science. In *Causation, Physics, and the Constitution of Reality: Russell's Republic Revisited*, ed. H. Price and R. Corry, 11–44. Oxford: Oxford University Press.

Nowak, M. A. 2006. Five rules for the evolution of cooperation. *Science* 314: 1560–1563.

Nowak, M. A., and K. Sigmund. 2005. Evolution of indirect reciprocity. *Nature* 437:1291–1298.

Nucci, L. 2001. *Education in the Moral Domain.* Cambridge: Cambridge University Press.

Oaksford, M., and N. Chater. 2007. *Bayesian Rationality: The Probabilistic Approach to Human Reasoning.* Oxford: Oxford University Press.

Onishi, K. H., and R. Baillargeon. 2005. Do 15-month-old infants understand false beliefs? *Science* 308:255–258.

Oppy, G. 2009. Cosmological arguments. *Noûs* 43:31–48.

Orians, G. H., and J. H. Heerwagen. 1992. Evolved responses to landscapes. In *The Adapted Mind: Evolutionary Psychology and the Generation of Culture*, ed. J. Barkow, L. Cosmides, and J. Tooby, 555–579. New York: Oxford University Press.

Otto, R. 1923. *The Idea of the Holy: An Inquiry into the Non-rational Factor in the Idea of the Divine and Its Relation to the Rational.* Trans. J. W. Harvey. Oxford: Oxford University Press.

Paley, W. (1802) 2006. *Natural Theology.* Ed. M. D. Eddy and D. Knight. Oxford: Oxford University Press.

Pannenberg, W. 2002. The concept of miracle. *Zygon: Journal of Religion and Science* 37:759–762.

Philipse, H. 2012. *God in the Age of Science? A Critique of Religious Reason.* Oxford: Oxford University Press.

Piaget, J. 1929. *The Child's Conception of the World.* Trans. J. Tomlinson and A. Tomlinson. London: Routledge and Kegan Paul.

Pike, A. W. G., D. L. Hoffmann, M. García-Diez, P. B. Pettitt, J. Alcolea, R. De Balbín, C. González-Sainz, et al. 2012. U-series dating of Paleolithic art in 11 caves in Spain. *Science* 336:1409–1413.

Pinker, S. 2002. *The Blank Slate: The Modern Denial of Human Nature.* New York: Viking.

Pitts, J. B. 2008. Why the Big Bang singularity does not help the kalām cosmological argument for theism. *British Journal for the Philosophy of Science* 59:675–708.

Plantinga, A. 1993. *Warrant and Proper Function.* Oxford: Oxford University Press.

Plantinga, A. 2000. *Warranted Christian Belief.* New York: Oxford University Press.

Plantinga, A. 2011. *Where the Conflict Really Lies: Science, Religion, and Naturalism.* Oxford: Oxford University Press.

Plato. (ca. 380 BCE) 2000. *Meno.* In *Exploring Philosophy: An Introductory Anthology,* ed. S. M. Cahn, 117–151. New York: Oxford University Press.

Polkinghorne, J. 1989. *Science and Providence: God's Interaction with the World.* London: Society for Promoting Christian Knowledge.

Polkinghorne, J. 1998. *Science and Theology: An Introduction.* Minneapolis: Fortress Press.

Pollard, B. 2005. Naturalizing the space of reasons. *International Journal of Philosophical Studies* 13:69–82.

Pollock, J. L. 1987. Defeasible reasoning. *Cognitive Science* 11:481–518.

Pope, M., K. Russel, and K. Watson. 2006. Biface form and structured behaviour in the Acheulean. *Lithics: The Journal of the Lithic Studies Society* 27:44–57.

Pospisil, L. J. 1978. *The Kapauku Papuans of West New Guinea.* New York: Holt, Rinehart & Winston.

Pouivet, R. 2011. Against theological fictionalism. *European Journal for Philosophy of Religion* 3:427–437.

Povinelli, D. J. 2000. *Folk physics for Apes: The Chimpanzee's Theory of How the World Works.* Oxford: Oxford University Press.

Povinelli, D. J., and S. Dunphy-Lelii. 2001. Do chimpanzees seek explanations? Preliminary comparative investigations. *Canadian Journal of Experimental Psychology* 55:185–193.

Prinz, J. 2006. The emotional basis of moral judgments. *Philosophical Explorations* 9:29–43.

Prum, R. O. 2012. Aesthetic evolution by mate choice: Darwin's *really* dangerous idea. *Philosophical Transactions of the Royal Society of London, Series B: Biological Sciences* 367:2253–2265.

Purzycki, B. G. 2013. The minds of gods: A comparative study of supernatural agency. *Cognition* 129:163–179.

Purzycki, B. G., D. N. Finkel, J. Shaver, N. Wales, A. B. Cohen, and R. Sosis. 2012. What does God know? Supernatural agents' access to socially strategic and nonstrategic information. *Cognitive Science* 36:846–869.

Pyysiäinen, I. 2002. Mind and miracles. *Zygon: Journal of Religion and Science* 37:729–740.

Pyysiäinen, I. 2003. True fiction: Philosophy and psychology of religious belief. *Philosophical Psychology* 16:109–125.

Pyysiäinen, I. 2009. *Supernatural Agents: Why We Believe in Souls, Gods, and Buddhas.* Oxford: Oxford University Press.

Pyysiäinen, I., M. Lindeman, and T. Honkela. 2003. Counterintuitiveness as the hallmark of religiosity. *Religion* 33:341–355.

Quinn, P. L. 2000. Divine command theory. In *The Blackwell Guide to Ethical Theory,* ed. H. LaFollette, 53–73. Malden, MA: Blackwell.

Qureshi, A., I. Apperly, and D. Samson. 2010. Executive function is necessary for perspective selection, not Level-1 visual perspective calculation: Evidence from a dual-task study of adults. *Cognition* 117:230–236.

Rahner, K. 1982. Theology and the arts. *Thought* 57:17–29.

Rasmussen, J. 2010. From states of affairs to a necessary being. *Philosophical Studies* 148:183–200.

Real, L. A. 1991. Animal choice behavior and the evolution of cognitive architecture. *Science* 253:980–986.

Reid, T. 1764. *An Inquiry into the Human Mind, on the Principles of Common Sense.* Edinburgh: Millar, Kincaid & Bell.

Richerson, P. J., and R. Boyd. 2005. *Not by Genes Alone: How Culture Transformed Human Evolution.* Chicago: University of Chicago Press.

Roes, F. L., and M. Raymond. 2003. Belief in moralizing gods. *Evolution and Human Behavior* 24:126–135.

Rogers, D. S., and P. R. Ehrlich. 2008. Natural selection and cultural rates of change. *Proceedings of the National Academy of Sciences of the United States of America* 105: 3416–3420.

Rogers, K. 2005. God and moral realism. *International Philosophical Quarterly* 45: 103–118.

Rossano, M. J. 2010. *Supernatural Selection: How Religion Evolved.* Oxford: Oxford University Press.

Rottman, J., and D. Kelemen. 2012. Is there such a thing as a Christian child? Evidence of religious beliefs in early childhood. In *Science and the World's Religions: Origins and Destinies,* ed. P. McNamara and W. Wildman, 205–238. Santa Barbara, CA: Praeger Press.

Rowe, W. L. 1979. The problem of evil and some varieties of atheism. *American Philosophical Quarterly* 16:335–341.

Rowe, W. L. 2005. Cosmological arguments. In *The Blackwell Guide to the Philosophy of Religion,* ed. W. E. Mann, 103–116. Malden, MA: Blackwell.

Runciman, W. G. 1999. Greek hoplites, warrior culture, and indirect bias. *Journal of the Royal Anthropological Institute* 4:731–751.

Ruse, M. 1995. *Evolutionary Naturalism: Selected Essays.* London: Routledge.

Ruse, M. 1998. Introduction. In *Philosophy of Biology,* ed. M. Kieran and D. McIver Lopes, 1–26. Amherst, NY: Prometheus Books.

Ruse, M. 2010. The biological sciences can act as a ground for ethics. In *Contemporary Debates in Philosophy of Biology,* ed. F. J. Ayala and R. Arp, 297–315. Chichester: Wiley-Blackwell.

Ruse, M., and E. O. Wilson. 1986. Moral philosophy as applied science. *Philosophy* 61:173–192.

Saler, B. 2008. Conceptualizing religion: Some recent reflections. *Religion* 38:219–225.

Samarapungavan, A., S. Vosniadou, and W. Brewer. 1996. Mental models of the Earth, Sun, and Moon: Indian children's cosmologies. *Cognitive Development* 11: 491–521.

Samarapungavan, A., and R. W. Wiers. 1997. Children's thoughts on the origin of species: A study of explanatory coherence. *Cognitive Science* 21:147–177.

Samson, D., and I. Apperly. 2010. There is more to mind reading than having theory of mind concepts: New directions in theory of mind research. *Infant and Child Development* 19:443–454.

Samuels, R. 2002. Nativism in cognitive science. *Mind and Language* 17:233–265.

Sanderson, S. K., and W. W. Roberts. 2008. The evolutionary forms of the religious life: A cross-cultural, quantitative analysis. *American Anthropologist* 110:454–466.

Sarkar, S. 2011. The science question in Intelligent Design. *Synthese* 178:291–305.

Saroglou, V., C. Buxant, and J. Tilquin. 2008. Positive emotions as leading to religion and spirituality. *Journal of Positive Psychology* 3:165–173.

Saxe, R., J. Tenenbaum, and S. A. Carey. 2005. Secret agents: Inferences about hidden causes by 10- and 12-month-old infants. *Psychological Science* 16:995–1001.

Schjoedt, U., H. Stødkilde-Jørgensen, A. W. Geertz, and A. Roepstorff. 2009. Highly religious participants recruit areas of social cognition in personal prayer. *Social Cognitive and Affective Neuroscience* 4:199–207.

Schloss, J. P., and M. Murray, eds. 2009. *The Believing Primate: Scientific, Philosophical, and Theological Reflections on the Origin of Religion.* Oxford: Oxford University Press.

Schloss, J. P., and M. Murray. 2011. Evolutionary accounts of belief in supernatural punishment: A critical review. *Religion, Brain and Behavior* 1:46–66.

Scholl, B. J., and P. D. Tremoulet. 2000. Perceptual causality and animacy. *Trends in Cognitive Sciences* 4:299–309.

Schulz, L. E., and J. Sommerville. 2006. God does not play dice: Causal determinism and preschoolers' causal inferences. *Child Development* 77:427–442.

Sedley, D. 2007. *Creationism and Its Critics in Antiquity.* Berkeley: University of California Press.

Sellars, W. 1956. Empiricism and the philosophy of mind. In *Minnesota Studies in the Philosophy of Science, Foundations of Science, and the Concepts of Psychology and Psychoanalysis,* ed. H. Feigl and M. Scriven, 253–329. Minneapolis, MN: University of Minnesota Press.

Semaw, S., P. Renne, J. W. K. Harris, C. S. Feibel, R. L. Bernor, N. Fesseha, and K. Mowbray. 1997. 2.5-million-year-old stone tools from Gona, Ethiopia. *Nature* 385:333–336.

Shanafelt, R. 2004. Magic, miracle, and marvels in anthropology. *Ethnos: Journal of Anthropology* 69:317–340.

Shariff, A. F., A. Norenzayan, and J. Henrich. 2010. The birth of high gods: How the cultural evolution of supernatural policing influenced the emergence of complex, cooperative human societies, paving the way for civilization. In *Evolution, Culture, and the Human Mind*, ed. M. Schaller, A. Norenzayan, S. J. Heine, T. Yamagishi, and T. Kameda, 179–195. New York: Psychology Press.

Sherman, P. W. 1977. Nepotism and the evolution of alarm calls. *Science* 197: 1246–1253.

Shihadeh, A. 2008. The existence of God. In *The Cambridge Companion to Classical Islamic Theology*, ed. T. Winter, 197–217. Cambridge: Cambridge University Press.

Shiota, M. N., D. Keltner, and A. Mossman. 2007. The nature of awe: Elicitors, appraisals, and effects on self-concept. *Cognition and Emotion* 21:944–963.

Shtulman, A. 2008. Variation in the anthropomorphization of supernatural beings and its implications for cognitive theories of religion. *Journal of Experimental Psychology: Learning, Memory, and Cognition* 34:1123–1138.

Shtulman, A., and J. Valcarcel. 2012. Scientific knowledge suppresses but does not supplant earlier intuitions. *Cognition* 124:209–215.

Shultz, T. R. 1982. Rules of causal attribution. *Monographs of the Society for Research in Child Development* 47:1–51.

Siegal, M., G. Butterworth, and P. A. Newcombe. 2004. Culture and children's cosmology. *Developmental Science* 7:308–324.

Silk, J. B., S. F. Brosnan, J. Vonk, J. Henrich, D. J. Povinelli, A. S. Richardson, S. P. Lambeth, J. Mascaro, and S. J. Schapiro. 2005. Chimpanzees are indifferent to the welfare of unrelated group members. *Nature* 437:1357–1359.

Silk, J. B., and B. R. House. 2011. Evolutionary foundations of human prosocial sentiments. *Proceedings of the National Academy of Sciences of the United States of America* 108:10910–10917.

Sinnott-Armstrong, W. 2009. Why traditional theism cannot provide an adequate foundation for morality. In *Is Goodness without God Good Enough? A Debate on Faith, Secularism, and Ethics*, ed. N. L. King and R. K. Garcia, 101–115. Lanham, MD: Rowman & Littlefield.

Slone, D. J. 2004. *Theological Incorrectness: Why Religious People Believe What They Shouldn't*. Oxford: Oxford University Press.

Smith, Q. 1999. The reason the universe exists is that it caused itself to exist. *Philosophy* 74:579–586.

Snow, J. 1855. *On the Mode of Communication of Cholera*, 2nd ed. London: John Churchill.

Sober, E. 2002. Intelligent design and probability reasoning. *International Journal for Philosophy of Religion* 52:65–80.

Sober, E. 2004. The design argument. In *The Blackwell Guide to the Philosophy of Religion*, ed. W. E. Mann, 117–147. Malden, MA: Blackwell.

Sommers, T., and A. Rosenberg. 2003. Darwin's nihilistic idea: Evolution and the meaninglessness of life. *Biology and Philosophy* 18:653–668.

Sosis, R., and C. Alcorta. 2003. Signaling, solidarity, and the sacred: The evolution of religious behavior. *Evolutionary Anthropology* 12:264–274.

Sosis, R., and E. R. Bressler. 2003. Cooperation and commune longevity: A test of the costly signaling theory of religion. *Cross-Cultural Research* 37:211–239.

Southgate, C. 2008. *The Groaning of Creation: God, Evolution, and the Problem of Evil.* Louisville, KY: Westminster John Knox Press.

Spelke, E. S. 1990. Principles of object perception. *Cognitive Science* 14:29–56.

Spelke, E. S., and K. D. Kinzler. 2007. Core knowledge. *Developmental Science* 10:89–96.

Spelke, E. S., A. Phillips, and A. L. Woodward. 1995. Infants' knowledge of object motion and human action. In *Causal Cognition: A Multidisciplinary Debate*, ed. D. Sperber, D. Premack, and A. J. Premack, 44–78. Oxford: Clarendon Press.

Sperber, D. 1985. Anthropology and psychology: Towards an epidemiology of representations. *Man* 20:73–89.

Sperber, D. 1996. *Explaining Culture: A Naturalistic Approach.* Oxford: Blackwell.

Sperber, D. 1997. Intuitive and reflective beliefs. *Mind and Language* 12:67–83.

Stark, R. 1999. Atheism, faith, and the social scientific study of religion. *Journal of Contemporary Religion* 14:41–61.

Steinbeis, N., and S. Koelsch. 2009. Understanding the intentions behind manmade products elicits neural activity in areas dedicated to mental state attribution. *Cerebral Cortex* 19:619–623.

Stephens, C. L. 2001. When is it selectively advantageous to have true beliefs? Sandwiching the better safe than sorry argument. *Philosophical Studies* 105:161–189.

Stewart-Williams, S. 2005. Innate ideas as a naturalistic source of metaphysical knowledge. *Biology and Philosophy* 20:791–814.

Stewart-Williams, S. 2010. *Darwin, God, and the Meaning of Life: How Evolutionary Theory Undermines Everything You Thought You Knew.* Cambridge: Cambridge University Press.

Street, S. 2006. A Darwinian dilemma for realist theories of value. *Philosophical Studies* 127:109–166.

Stump, E. 1997. Awe and atheism. *Midwest Studies in Philosophy* 21:281–289.

Subramuniyaswami, S. S. 2000. *Loving Ganeśa: Hinduism's Endearing Elephant-Faced God*. Kapaa, HI: Himalayan Academy.

Sudduth, M. C. 1995. The prospects for "mediate" natural theology in John Calvin. *Religious Studies* 31:53–68.

Sullivan, P. R. 2009. Objects limit human comprehension. *Biology and Philosophy* 24:65–79.

Sundararajan, L. 2002. Religious awe: Potential contributions of negative theology to psychology, "positive" or otherwise. *Journal of Theoretical and Philosophical Psychology* 22:174–197.

Surian, L., S. Caldi, and D. Sperber. 2007. Attribution of beliefs by 13-month-old infants. *Psychological Science* 18:580–586.

Sutton, J. 2005. Stick to what you know. *Noûs* 39:359–396.

Swinburne, R. 1968a. The argument from design. *Philosophy* 43:199–212.

Swinburne, R. 1968b. Miracles. *Philosophical Quarterly* 18:320–328.

Swinburne, R. 2004. *The Existence of God*, 6th ed. Oxford: Clarendon Press.

Swinburne, R. 2010. *Is There a God?* rev. ed. Oxford: Oxford University Press.

Taliaferro, C. 2005. *Evidence and Faith: Philosophy and Religion since the Seventeenth Century*. Cambridge: Cambridge University Press.

Taylor, J. E. 2007. Hume on miracles: Interpretation and criticism. *Philosophy Compass* 2:611–624.

Taylor, M., B. Esbensen, and R. Bennett. 1994. Children's understanding of knowledge acquisition: The tendency for children to report that they have always known what they have just learned. *Child Development* 65:1581–1604.

Teilhard de Chardin, P. 1959. *The Phenomenon of Man*. Trans. B. Wall. New York: Harper & Row.

Tennant, F. R. 1930. *Philosophical Theology: The World, the Soul, and God*, vol. 2. Cambridge: Cambridge University Press.

Thurston, W. 2006. On proof and progress in mathematics. In *18 Unconventional Essays on the Nature of Mathematics*, ed. R. Hersh, 37–55. New York: Springer.

Tillich, P. 1953. *Systematic Theology*, vol. 1. London: James Nisbet.

Tomasello, M. 1999. *The Cultural Origins of Human Cognition*. Cambridge, MA: Harvard University Press.

Trollope, A. (1857) 1994. *Barchester Towers*. Ware: Wordsworth Classics.

Tylor, E. B. 1871. *Primitive Culture: Researches into the Development of Mythology, Philosophy, Religion, Language, Art, and Custom*. London: John Murray.

Ulrich, R. S. 1984. View through a window may influence recovery from surgery. *Science* 224:420–421.

Ulrich, R. S. 1993. Biophilia, biophobia, and natural landscapes. In *The Biophilia Hypothesis*, ed. S. R. Kellert and E. O. Wilson, 73–137. Washington, D.C.: Island Press.

Valdesolo, P., and J. Graham. 2014. Awe, uncertainty, and agency detection. *Psychological Science* 25:170–178.

Valladas, H. 2003. Direct radiocarbon dating of prehistoric cave paintings by accelerator mass spectrometry. *Measurement Science and Technology* 14:1487–1492.

Van Damme, W. 1997. Do non-western cultures have words for art? An epistemological prolegomenon to the comparative study of philosophies of art. In *Proceedings of the Pacific Rim conference in transcultural aesthetics*, ed. E. Benitez, 97–115. Sydney: University of Sydney.

van Fraassen, B. C. 1989. *Laws and Symmetry*. Oxford: Clarendon.

Vanhaeren, M., and F. d'Errico. 2005. Grave goods from the Saint-Germain-la-Rivière burial: Evidence for social inequality in the Upper Palaeolithic. *Journal of Anthropological Archaeology* 24:117–134.

Vanhaeren, M., F. d'Errico, C. Stringer, S. James, J. Todd, and H. Mienis. 2006. Middle Paleolithic shell beads in Israel and Algeria. *Science* 312:1785–1788.

van Inwagen, P. 1999. Is it wrong everywhere, always, and for anyone to believe anything on insufficient evidence? In *Philosophy of Religion: The Big Questions*, ed. E. Stump and M. Murray, 273–284. Malden: Blackwell.

van Inwagen, P. 2006. *The Problem of Evil*. Oxford: Clarendon.

van Woudenberg, R. 2005. Intuitive knowledge reconsidered. In *Basic Belief and Basic Knowledge: Papers in Epistemology*, ed. R. van Woudenbergh, S. Roeser, and R. Rood, 15–39. Heusenstamm nr. Frankfurt: Ontos Verlag.

Vartanian, O., and V. Goel. 2004. Neuroanatomical correlates of aesthetic preference for paintings. *Neuroreport* 15:893–897.

Verweij, J., P. Ester, and R. Nauta. 1997. Secularization as an economic and cultural phenomenon: A cross-national analysis. *Journal for the Scientific Study of Religion* 36:309–324.

Viladesau, R. 2008. *Theosis* and beauty. *Theology Today* 65:180–190.

Visala, A. 2011. *Naturalism, Theism, and the Cognitive Study of Religion: Religion Explained?* Farnham: Ashgate.

Vonk, J., and D. J. Povinelli. 2006. Similarity and difference in the conceptual systems of primates: The unobservability hypothesis. In *Comparative Cognition: Experimental Explorations of Animal Intelligence*, ed. E. Wasserman and T. Zentall, 363–387. Oxford: Oxford University Press.

Wallace, A. R. 1858. On the tendency of varieties to depart indefinitely from the original type. *Proceedings of the Linnean Society of London* 3:53–62.

Wallace, A. R. 1870. *Contributions to the Theory of Natural Selection: A Series of Essays.* London: Macmillan.

Wallace, A. R. 1877. The colours of animals and plants, I. The colours of animals. *Macmillan's Magazine* 36:384–408.

Ward, K. 1985. Miracles and testimony. *Religious Studies* 21:131–145.

Ward, K. 2002. Believing in miracles. *Zygon: Journal of Religion and Science* 37:741–750.

Wegener, A. 1912. Die Entstehung der Kontinente. *Geologische Rundschau. Zeitschrift für allgemeine Geologie* 3:276–292.

Wegner, D. M. 2003. The mind's self-portrait. *Annals of the New York Academy of Sciences* 1001:212–225.

Wellman, H. M., and D. Liu. 2004. Scaling of theory-of-mind tasks. *Child Development* 75:523–541.

Wenger, A., and B. J. Fowers. 2008. Positive illusions in parenting: Every child is above average. *Journal of Applied Social Psychology* 3:611–634.

Wenger, J. L. 2001. Children's theories of God: Explanations for difficult-to-explain phenomena. *Journal of Genetic Psychology* 162:41–55.

West, S. A., and A. Gardner. 2010. Altruism, spite, and greenbeards. *Science* 327:1341–1344.

Wettstein, H. 2012. *The Significance of Religious Experience.* Oxford: Oxford University Press.

Whallon, R. 2006. Social networks and information: Non-utilitarian mobility among hunter-gatherers. *Journal of Anthropological Archaeology* 25:259–270.

Whitehouse, H. 2004. *Modes of Religiosity: A Cognitive Theory of Religious Transmission.* Walnut Creek, CA: AltaMira Press.

Whiten, A., J. Goodall, W. C. McGrew, T. Nishida, V. Reynolds, Y. Sugiyama, C. E. G. Tutin, R. W. Wrangham, and C. Boesch. 1999. Cultures in chimpanzees. *Nature* 399:682–685.

Wierenga, E. 2009. Omniscience. In *The Oxford Handbook of Philosophical Theology*, ed. T. P. Flint and M. C. Rea, 129–144. Oxford: Oxford University Press.

Wilkins, J. S., and P. E. Griffiths. 2013. Evolutionary debunking arguments in three domains: Fact, value, and religion. In *A New Science of Religion*, ed. G. W. Dawes and J. Maclaurin, 133–146. New York: Routledge.

Williamson, T. 2007. *The Philosophy of Philosophy*. Oxford: Blackwell.

Wilson, E. O. 1984. *Biophilia*. Cambridge, MA: Harvard University Press.

Winner, L. 2006. *Girl Meets God: On the Path to a Spiritual Life*. Milton Keynes: Authentic Media.

Wolterstorff, N. 1983. Can belief in God be rational if it has no foundations? In *Faith and Rationality: Reason and Belief in God*, ed. A. Plantinga and N. Wolterstorff, 135–186. Notre Dame, IN: University of Notre Dame Press.

Wynn, K. 1992. Addition and subtraction by human infants. *Nature* 358:749–750.

Wynn, M. 1997. Beauty, providence and the biophilia hypothesis. *Heythrop Journal* 38:283–299.

Xenophon. (4th c. BCE) 1997. Memorabilia. In *Memorabilia, Oeconomicus, Symposium, Apology*, trans. E. C. Marchant, and O. J. Todd, 1–359. Cambridge, MA: Harvard University Press.

Young, L., and A. J. Durwin. 2013. Moral realism as moral motivation: The impact of meta-ethics on everyday decision-making. *Journal of Experimental Social Psychology* 49:302–306.

Zagzebski, L. 2008. Omnisubjectivity. In *Oxford Studies in Philosophy of Religion*, vol. 1, ed. J. L. Kvanvig, 231–248. Oxford: Oxford University Press.

Zahavi, A. 1975. Mate selection: A selection for a handicap. *Journal of Theoretical Biology* 53:205–214.

Zeki, S. 1999. *Inner Vision*. New York: Oxford University Press.

Zuckerman, P. 2007. Atheism: Contemporary numbers and patterns. In *The Cambridge Companion to Atheism*, ed. M. Martin, 47–65. Cambridge: Cambridge University Press.

Index

Printed in the United States
by Baker & Taylor Publisher Services